Providing a concise and accessible introduction to the work of the celebrated twentieth century German philosopher, Hans-Georg Gadamer, this book focuses on the aspects of Gadamer's philosophy that have been the most influential among architects, educators in architecture, and architectural theorists.

Gadamer's philosophy of art gives a special place to the activity of "play" as it occurs in artistic creation. His reflections on meaning and symbolism in art draw upon his teacher, Martin Heidegger, while moving Heidegger's thought in new directions. His theory of interpretation, or "philosophical hermeneutics," offers profound ways to understand the influence of the past upon the present and to appropriate cultural history in ever newer forms.

For architects, architectural theorists, architectural historians, and students in these fields, Gadamer's thought opens a world of possibilities for understanding how building today can be rich with human meaning, relating to architecture's history in ways that do not merely repeat nor merely repudiate that history.

In addition, Gadamer's sensitivity to the importance of practical thinking—to the way that theory arises out of practice—gives his thought a remarkable usefulness in the everyday work of professional life.

Paul Kidder is Associate Professor of Philosophy at Seattle University, where he has taught courses on ethics, urban life, and the philosophy of art.

Thinkers for Architects

Series Editor: Adam Sharr, Newcastle University, UK

Editorial Board
Jonathan A. Hale, University of Nottingham, UK
Hilde Heynen, KU Leuven, Netherlands
David Leatherbarrow, University of Pennsylvania, USA

Architects have often looked to philosophers and theorists from beyond the discipline for design inspiration or in search of a critical framework for practice. This original series offers quick, clear introductions to key thinkers who have written about architecture and whose work can yield insights for designers.

> "Each unintimidatingly slim book makes sense of the subjects' complex theories."
>
> Building Design

> "... a valuable addition to any studio space or computer lab"
>
> Architectural Record

> "... a creditable attempt to present their subjects in a useful way"
>
> Architectural Review

Deleuze and Guattari for Architects
Andrew Ballantyne

Heidegger for Architects
Adam Sharr

Irigaray for Architects
Peg Rawes

Bhabha for Architects
Felipe Hernández

Bourdieu for Architects
Helena Webster

Benjamin for Architects
Brian Elliott

Derrida for Architects
Richard Coyne

Gadamer for Architects
Paul Kidder

Foucault for Architects
Gordana Fontana-Giusti

THINKERS FOR ARCHITECTS

Gadamer
for
Architects

Paul Kidder

LONDON AND NEW YORK

First published 2013
by Routledge
2 Park Square, Milton Park, Abingdon, Oxon OX14 4RN

Simultaneously published in the USA and Canada
by Routledge
711 Third Avenue, New York, NY 10017

Routledge is an imprint of the Taylor & Francis Group, an informa business

© 2013 Paul Kidder

The right of Paul Kidder to be identified as author of this work has been asserted by him/her in accordance with sections 77 and 78 of the Copyright, Designs and Patents Act 1988.

All rights reserved. No part of this book may be reprinted or reproduced or utilised in any form or by any electronic, mechanical, or other means, now known or hereafter invented, including photocopying and recording, or in any information storage or retrieval system, without permission in writing from the publishers.

Trademark notice: Product or corporate names may be trademarks or registered trademarks, and are used only for identification and explanation without intent to infringe.

British Library Cataloguing-in-Publication Data
A catalogue record for this book is available from the British Library

Library of Congress Cataloging in Publication Data
Kidder, Paul.
 Gadamer for architects / Paul Kidder.
   pages cm -- (Thinkers for architects)
 Includes bibliographical references and index.
 1. Gadamer, Hans-Georg, 1900–2002. 2. Architecture--Philosophy.
 I. Title.
 B3248.G34K53 2012
 193--dc23
                                                          2012027345

ISBN: 978-0-415-52272-4 (hbk)
ISBN: 978-0-415-52273-1 (pbk)
ISBN: 978-0-203-07322-3 (ebk)

Typeset in Frutiger and Galliard
by Fakenham Prepress Solutions, Fakenham, Norfolk NR21 8NN

Printed and bound in Great Britain by
TJ International Ltd, Padstow, Cornwall

For Paulette

Contents

Series Editor's Preface   xi
Acknowledgements   xiii
Illustration Credits   xiii

1 Introduction   1

    "Hermeneutics" and "hermeneutic" 2
    Philosophical hermeneutics and the aims of architecture 5
    Gadamer in the context of European thought 10
    Plan of the book 15

2 The Play of Art and the Art of Architecture   17

    The problem of the subjectivization of art 18
    Art and "serious play" 21
    The speechless image and the embodied word 22
    The "decorative" nature of architecture 28
    Questions of balance 32
    Play as clue and instance 34

3 Historical Understanding and Architecture's Past   38

    The idea of horizon 39
    The interpretation of texts 43
    Horizon and history 46
    Hermeneutics and history in architectural theory 50
    The hermeneutics of sacred architecture 54
    Architecture's past in present design 56

4 Humanism in the Age of Science   64

   Humanity in the emergence of the natural-scientific ideal 66
   Modernity and the ethical function of architecture 69
   Hermeneutics and the rationalization of society 72
   Social engineering and the city as work of art 76

5 Practical Wisdom in Creative Collaboration   81

   The practical wisdom tradition 81
   Hermeneutics in practical deliberation 84
   Collaboration in architectural creation 87
   Design as horizon 89
   The example of the rural studio 92
   Finding measure 95

6 Architecture as a Way of Being   97

   Some puzzles about being 97
   Heideggerian thinking of being 98
   Hermeneutics and Heideggerian ontology 101
   The ontology of building and place 103
   The ontology of language 106
   The ontology of time 107
   The hermeneutics of suspicion 109
   Transcendence and mystery 116

   Conclusion: The Hermeneutically Minded Architect   120
   For Further Reading   124
   Bibliography   126
   Index   136

Series Editor's Preface

Adam Sharr

Architects have often looked to thinkers in philosophy and theory for design ideas, or in search of a critical framework for practice. Yet architects and students of architecture can struggle to navigate thinkers' writings. It can be daunting to approach original texts with little appreciation of their contexts. And existing introductions seldom explore architectural material in any detail. This original series offers clear, quick and accurate introductions to key thinkers who have written about architecture. Each book summarizes what a thinker has to offer for architects. It locates their architectural thinking in the body of their work, introduces significant books and essays, helps decode terms and provides quick reference for further reading. If you find philosophical and theoretical writing about architecture difficult, or just don't know where to begin, this series will be indispensable.

Books in the *Thinkers for Architects* series come out of architecture. They pursue architectural modes of understanding, aiming to introduce a thinker to an architectural audience. Each thinker has a unique and distinctive ethos, and the structure of each book derives from the character at its focus. The thinkers explored are prodigious writers and any short introduction can only address a fraction of their work. Each author – an architect or an architectural critic – has focused on a selection of a thinker's writings which they judge most relevant to designers and interpreters of architecture. Inevitably, much will be left out. These books will be the first point of reference, rather than the last word, about a particular thinker for architects. It is hoped that they will encourage you to read further; offering an incentive to delve deeper into the original writings of a particular thinker.

The Thinkers for Architects series has proved highly successful, expanding now to eight volumes dealing with familiar cultural figures whose writings have influenced architectural designers, critics and commentators in distinctive and important ways. Books explore the work of: Gilles Deleuze and Felix Guattari; Martin Heidegger; Luce Irigaray; Homi Bhabha; Pierre Bourdieu, Walter Benjamin; Jacques Derrida and Hans-Georg Gadamer. A number of future volumes are projected, addressing Maurice Merleau-Ponty, Michel Foucault and Jean Baudrillard. The series continues to expand, addressing an increasingly rich diversity of contemporary thinkers who have something to say to architects.

Adam Sharr is Professor of Architecture at the University of Newcastle-upon-Tyne, Principal of Adam Sharr Architects and Editor (with Richard Weston) of *arq: Architectural Research Quarterly*, Cambridge University Press' international architecture journal. His books include *Heidegger for Architects*.

Acknowledgements

The idea for this book emerged out of my participation in the 2011 Architecture+Philosophy conference at Boston University. I am grateful to Bryan Norwood, Elizabeth Robinson, Daniel Dahlstrom, and other organizers of that conference for the opportunity to share and develop ideas that are presented here. I am grateful to series editor Adam Sharr and the editors and reviewers for Routledge who encouraged this project at every step, including Laura Williamson, Georgina Johnson, and Francesca Ford. I give thanks to the many teachers, colleagues, and friends who have sparked, supported, tutored, or indulged my interest in architecture, hermeneutics, or both over the years, including Richard Cobb-Stevens, Charles Lawrence, Frederick Lawrence, Thomas McPartland, Fr. William Richardson, and James Risser. It was long ago that the late Fr. Joseph Flanagan simply insisted that I do work on the philosophy of architecture, and the book is a product of that insistence. It is the fruit, as well, of Professor Gadamer's endless patience, in those same years, with the many questions of a young student lost in a forest of ideas. Glenn Hughes is to be deeply thanked for being, during most of my adult life, my chief mentor and interlocutor on matters philosophical, poetic, and artistic. The book was prepared with support from the faculty development program of the Seattle University College of Arts and Sciences and its Deans, Wallace Loh and David Powers; the support of the Department of Philosophy and its Chair, Burt Hopkins; and the encouragement of my family. To Paulette Kidder, also a Gadamer scholar and my companion in all things, the book is gratefully dedicated.

Illustration credits

Cottage at Niarbyl, near Peel, Isle of Man. Photo by David J. Radcliffe.
Frank Gehry, Disney Concert Hall, Los Angeles. Photos by Paul Kidder.

Frank Gehry, Ray and Maria Stata Center, Massachusetts Institute of Technology, Cambridge, Massachusetts. Photo by Paul Kidder.

Adolf Loos, "Looshaus," Michaelerplatz, Vienna. Photo by Andreas Praefcke.

H. H. Richardson, Trinity Church, and Henry N. Cobb for I. M. Pei & Partners, John Hancock Tower, Boston. Photo by Paul Kidder.

Michael Graves, Portland Building, Portland, Oregon. Photo by Paul Kidder.

Steven Holl, Chapel of St. Ignatius, Seattle. Photos by Paul Kidder.

North End, Boston. Photo by Paul Kidder.

Rural Studio, Bryant "Hay Bale" House, Mason's Bend, Hale County, Alabama. Photo by Kirby Davis.

CHAPTER 1

Introduction

Hans-Georg Gadamer was one of the most prominent European thinkers of the twentieth century. His philosophy is above all dedicated to discovering the most general patterns of experience and thinking that occur whenever people seek to understand the world and one another, whenever they interpret texts or other expressions of meaning, and whenever they experience art or nature as intriguing, enjoyable, and significant. The pattern that Gadamer came to see running through all of these forms of experience he called "hermeneutic," and he characterized his philosophy as "philosophical hermeneutics." Today in many fields of study one can find theorists who take a "hermeneutic" approach to their subject, and in most such cases the influence of Gadamer can be recognized.

While Gadamer wrote little on the subject of architecture specifically, his writings on art and aesthetics are extensive, and they play a central role in his overall philosophical program. For this reason his thinking has, for decades, been influential on a variety of architects and architectural theorists. A hermeneutic approach to architecture discovers the hermeneutic pattern of thought and experience in several areas of the architectural enterprise. It may be found in the creative activity of the architect and the aesthetic appreciation of the architect's creations. It may be found in the way architects seek to understand architectural traditions and writings on architecture. It may be found in the ways that all of those who collaborate in the architectural enterprise—the clients, the community, the developers, the regulators, and the designers—understand one another and work together. It is important to note that in all such cases the hermeneutic pattern is, Gadamer would say, *inevitably* at work. It is something that happens whether or not we intend it. But by realizing and acknowledging that the pattern is at work we may consciously and deliberately seek to follow the path that it opens for us.

"Hermeneutics" and "hermeneutic"

The root of the word "hermeneutics" is an ancient one referring to the activity of interpretation. Already in the philosophies of Plato and Aristotle, and in the writings of ancient rhetoricians, there occurred the question as to how people understand and misunderstand one another, and how words that are written down can be interpreted when the author is not present to be questioned in dialogue. In the eighteenth and nineteenth centuries in Europe, "hermeneutics" became an important subfield within the disciplines of law, theology, and modern rhetoric. Theorists in these fields sought to develop rigorous standards and techniques for determining how texts that were written in previous ages should be applied in contemporary circumstances. Religious scriptures and legal statutes, for example, may have been formulated in a very different time, yet they intend that their readers should live by them today. The documents are not merely of historical interest, but make real claims upon their readership. Hence the work of hermeneutics in these disciplines always combines understanding with practical decision and action (Gadamer 1990b, 324–34).

Gadamer claims that this quality of traditional hermeneutics—its combination of intellectual grasp and practical application—is at work in the interpretation of any text, and indeed in every instance of trying to understand the thoughts and beliefs of another person or culture. He holds that the hermeneutic dimension of understanding is universal (1976, 3–17). Architecture, like texts, always functions, to some degree, as a carrier of cultural meaning. Dalibor Vesely has said this in a dramatic way: "what the book is to literacy," he writes, "architecture is to culture as a whole" (Vesely 2004, 8). If there is an inevitable pattern of understanding that occurs whenever a work of architecture is interpreted, should the architect not have a sense of how that pattern unfolds? If architecture itself interprets a culture, would it not benefit the architect to examine the pattern by which architecture forms such an interpretation? If buildings from the past, like books from the past, have a continued relevance for the present, should the architect not ponder the pattern by which that relevance is achieved? Hermeneutics, in both the study of texts and in relation to architecture, holds the promise of fundamentally altering the way one thinks about interpretation, understanding, and the communication of culture.

Hermeneutics holds the promise of fundamentally altering the way one thinks about interpretation, understanding, and the communication of culture.

The hermeneutic pattern that Gadamer seeks to articulate has been called "the hermeneutic circle." This idea does not originate with Gadamer. Already in the hermeneutics of the nineteenth century (which Gadamer calls "Romantic hermeneutics") there was a sense that the interpretation of texts must move in a circular pattern. This circle is often characterized in terms of a relationship of parts and wholes: to understand the whole of a book it is necessary to grasp its individual words and sentences, but those words and sentences only have meaning within the larger context of the book, hence interpretation must be a matter of constant revision: revising one's sense of the whole as one grasps the individual parts, and revising one's sense of the parts as the meaning of the whole emerges. Thus the hermeneutic circle is not a vicious circle but a cumulatively productive one (1988, 68–78; 1990b, 190–91).

There is another, more subtle, way in which a hermeneutic circle is at work in such cases. When the philosopher and theologian Augustine (354–430) explained how one should study the Bible, he recommended interpreting the more obvious passages first and then going on to make sense of the difficult and obscure passages in terms of the obvious ones (Augustine 1958, 42–3). But practitioners of Romantic hermeneutics realized that what makes a passage obvious to a reader might be less a matter of its inherent transparency than a matter of the assumptions that the reader brings to the text. Moreover, because the passage seems obvious, one might be less inclined to question those assumptions than one might otherwise be. So the hermeneutic circle, in this light, must proceed by beginning with assumptions, but must then revise them as an understanding of the text deepens. Ultimately, one may come to see all of the obvious readings with which one began as misinterpretations born of naïveté and bias (Gadamer 1990b, 179–80; Schleiermacher, 1998).

Gadamer's major philosophical work is entitled *Truth and Method*. The first word in this title is meant to suggest that the hermeneutic pattern is involved whenever one seeks truth. Here Gadamer deliberately echoes Plato. Plato used the word "dialectic" to refer to the general form of Socratic inquiry that questions commonsense assumptions, seeks consistent definitions of terms, considers explanatory hypotheses, formulates arguments, considers objections, and refutes those objections or revises the theory in light of them. Dialectic can be undertaken with regard to any question or any topic, be it a question of nature, of morality, of politics, of religion, or of art. It can be pursued in good-natured discussion among friends, or it can be practiced in the privacy of one's own reflections (Gadamer 1980, 93–123; 1990b, 362–9). Gadamer advocated the same kind of generality for his hermeneutic pattern that Plato claimed for dialectic. To make this point, Gadamer sometimes proposed dropping the "s" off of the end of the English word "hermeneutics" to show that "hermeneutic" has the same comprehensiveness and generality as dialectic, and goes together with dialectic.

"Method," on the other hand, is something more modern and narrower in scope than either hermeneutic or dialectic. The notion of "method," as it is commonly used today, was born of the eighteenth-century Enlightenment period, when thinkers were seeking to associate rationality with the investigative techniques of natural science. A "method," in this context, identifies a limited set of questions to be asked, a limited range of evidence to be consulted, and a strict empirical procedure to be followed. A Newtonian discussion of the interaction of a set of billiard balls, for example, would be limited to questions of the physical forces by which they move and are moved; the evidence would be the observable interactions of the individual balls; the procedure might involve experiments that would isolate one kind of impact of one ball on another ball. Nothing would be said about the human interests of the people manipulating the cues or the about the game in which they are engaged. Never would the ultimate meaning of the movement of the balls or the value of the game come into consideration.

This narrowed form of investigation is of incalculable value within the limited scope of natural-scientific questions. The problem, however, is that Enlightenment culture began promoting this kind of rationality as the premier form while demoting all other forms. In the centuries since, as ever-increasing numbers of disciplines have sought to become "scientific," it is not uncommon to hear natural-scientific rationality proclaimed as the only legitimate form of rationality. Gadamer, in contrast, resists all such narrowing of the ideal of reason. For most of human history there has been a belief that the arts and humanities promote crucial forms of rationality, forms that civilize us and justify our ethics, our laws, and our institutions of government. To denigrate or to lose those forms of rationality is to risk a dehumanized society. It is for this reason that philosophical hermeneutics aims at recovering, in a contemporary manner, a more generalized and broadly applicable notion of reasoning and its uses. The search for truth, Gadamer's book title intends to suggest, is in some degree of tension with the idea of method (1984c, 151–69; 1990b, 3–9).

The search for truth, Gadamer's book title intends to suggest, is in some degree of tension with the idea of method.

Philosophical hermeneutics and the aims of architecture

As the book is to literacy, says Vesely, so architecture is to culture. This cultural significance of architecture is what attracts many students to the field. They have a sense that architecture is not simply interesting or enjoyable, but is important to life. It can contribute, in its own way, to enriching a community and expressing that community's significance. This dimension of meaning is part of what gives architecture an affinity to philosophy, but one need not be a student of either philosophy or architecture to understand, at some level, that architecture carries profound human meaning. Countless tourists board airplanes every year to be awed by the world's great buildings, gardens, and cities. In their visits to palaces, temples, and pyramids these people have a sense that they are not just encountering works of art, but expressions of the deepest cultural aspirations of a people. The great works of architecture compound many kinds of meaning and purpose into a single creation, and do so in an astoundingly integral way.

But how is architecture able to accomplish this? And how is it able to continue doing so through historical changes in style, technique, and materials? Such questions embroil us a complex set of issues within the field. Let us approach them simply by identifying four basic puzzles that characterize the discussion. A first of these would concern the matter of functionality. Architecture must be functional, and yet it should not be reducible to its functionality. A kitchen appliance may be called "wonderful," and even "the perfect combination of utility and design," but a building, even if it is a "machine for living," is a disappointment if it is only wonderful in the ways that an appliance is wonderful. A building should signify more than that. This is not to say that buildings must be monumental or overtly representational. A humble structure can be highly communicative; an abstract style can be rich with suggestiveness. But the architect always faces the challenge of making a structure meaningful, or of relating the structure to traditions of meaningful building, while at the same time fulfilling contemporary expectations for functionality under the constraints of a budget.

Secondly, while architecture is certainly an art, its cultural significance involves more than its aesthetics. For most of human history this point has seemed fairly obvious. The beauty of a Gothic arch—the way it fashions stone into elegant and dramatic shapes—is not separable from its structural function of holding up the roof—nor, for that matter, from the symbolic meaning it communicates by pointing to heaven and making heavy stone seem to evaporate into the air. All of these qualities are thoroughly intertwined. But a number of writers on the history of architecture (and notably several who are familiar with Gadamer) identify trends beginning in the Baroque period that separate the aesthetic dimension from the structural underpinning. The beauty of the work, the way it pleases the senses, comes to be associated with ornament, and the structure becomes mere scaffolding for ornament. The first part of Gadamer's *Truth and Method* gives an account of how this attenuated sense of the aesthetic came to dominate art criticism and the philosophy of art. The trend, for Gadamer, represents a weakening of the sense of the power and importance of art and architecture. Karsten Harries and Alberto Pérez-Gómez, drawing on Martin Heidegger and others in the European philosophical tradition, approach the issue in terms of architecture's ability to embody a cultural ethos. The aestheticization of architecture's sensuous qualities would seem to impair its ability to fulfill this "ethical" function (Harries 1997; Pérez-Gómez 2008, ch. 9; Sharr 2007, 101–3). Vesely speaks, in a similar vein, of the division of architectural representation from its technological support as a conundrum for contemporary architects, something calling for a new integration (2004, chs. 4 and 5; Sharr 2007, 103–4).

The modern innovations of figures such as Adolf Loos, Walter Gropius, Mies van der Rohe, Le Corbusier, and Louis Kahn represent daring efforts to reintegrate architecture in a radically new way, decrying superfluous ornament and seeking to draw out of the potentialities of new materials and engineering principles a meaningful vision of a thoroughly modern human community. But the realization of this modern vision has not gone entirely smoothly, with backlashes that include charges of social engineering, demands for historic preservation, and the emergence of various forms of "post"-modernism. It would be fair to say that the question persists as to how the aesthetic dimension can be reintegrated in a fully successful way with the other important qualities of architecture.

> The question persists as to how the aesthetic dimension can be reintegrated in a fully successful way with the other important qualities of architecture.

A third kind of puzzle has to do with the symbolic function of architecture. When we speak of the "meaning" of architecture, or when we follow, say, the approach of Christian Norberg-Schulz in his classic survey *Meaning in Western Architecture* (1975), what is intended by the word "meaning" is primarily the symbolic dimension of architectural works. The most primordial of structures can possess commanding symbolic suggestiveness. The monolith or the stone circle may employ the most elementary of building technologies, and yet they align with the stars and track the seasons in ways that seem to connect the structures with the whole of the cosmos. To create settlements in a landscape is always also to symbolize the human relationship to that landscape. Many forms of local and traditional architecture, for example, identify with the environment by building with the materials that it provides, orienting the resulting dwellings to the topography, the light, and the climate patterns of the region. As human construction identifies with landscape, it simultaneously complements the landscape by symbolizing the ways that human life stands out from nature. The monolith and the tower, for example, announce, by their rising-up from the ground and their orientation to the sky, the human way of standing-out from the world of minerals, plants, and non-human animals. The civic gateway and the domestic threshold (to take another simple example) divide, both practically and symbolically, varieties of public space from the interiority of community, family, or the individual psyche (cf. Norberg-Schulz 1979; 1985).

Cottage at Niarbyl, near Peel, Isle of Man

Countless symbolisms of this sort can be elaborated, yet a host of questions accompanies the task. What, one may ask, does it mean, after all, to be a symbol? How does symbolism embody meaning? Of the many kinds of scholars who study symbolism—psychologists, philosophers, theologians, anthropologists, historians—who is best equipped to explain the symbols of architecture? Do the architectural forms that carry symbolism need to be representational? Do they need to be part of a narrative? If so, does this mean that a more abstract kind of architecture is also a less symbolic kind? And what about those architects who go about their work without giving a thought to symbolism? Is their work free of symbolisms or do they symbolize in spite of themselves?

In the fourth place, all of the features that I have been describing—the functionality, the artistry, and the symbolism—are shaped within history and traditions. In "traditional" forms of architecture this fact can be expressed in a straightforward manner: the building solutions and stylistic tastes were preserved and handed down as the legacy of a tradition. But to the extent that modernism made its project the overcoming of the limitations of the past, it raised new questions as to how modern styles should relate to past styles. The nature of modernism is not simply to invent new artistic conventions for a new age, but to keep endlessly innovating. An *avant garde* will constantly be reinventing everything, and the most prominent architects today are under pressure to keep reinventing themselves—as Picasso regularly did—resisting the repetition of even their own former creations. Yet at the same time there are those who consider the idea of stepping out of history and forming an utterly new age or a new humanity as regrettably naïve. The past has a power over us, it is argued, that runs deeper than our abilities to consciously master it, and to reject our origins in a totalizing way is to leave ourselves alienated and homeless. This is the dilemma of history, within which every architect must take a stand.

Gadamer's hermeneutics is relevant to all four of these areas of questions and challenges for architecture. His efforts to recover ideals of rationality that are broader than empirical and technical reason parallel efforts in architectural theory to resist the reduction of architecture's significance to its practical functioning. The distinctive understanding of the notion of "play" that Gadamer develops in the first chapters of *Truth and Method* goes a long way toward integrating artistic activity with other forms of meaning-making, and particularly the search for truth. In a number of his writings Gadamer gives attention to the distinctive nature of symbolism when it occurs in the context of non-verbal forms of art. His philosophy, finally, devotes much attention to questions of the role of history and tradition in the unfolding of contemporary art and culture.

The past has a power over us, it is argued, that runs deeper than our abilities to consciously master it, and to reject our origins in a totalizing way is to leave ourselves alienated and homeless.

Gadamer in the context of European thought

Gadamer was a thinker in the German philosophical tradition. He is often described as the inheritor of the legacy of great thinkers that includes Immanuel Kant, G. W. F. Hegel, Edmund Husserl, and Martin Heidegger. Gadamer, moreover, tends to assume that his readers are familiar with that tradition. His writing style, he always insisted, is a good German style, and one that always sought to reflect qualities of dialogical speech. But for English readers, even when reading excellent translations, there are features of this style that can be daunting. German style is often "periodic" —that is, clauses and parenthetical points are built up within sentences and paragraphs in a way that delays the indication of the overall point of the passage. Gadamer's use of this style of writing can be especially baffling for readers who are not practiced in German literature or philosophy. Still, this challenge has not kept a great many non-specialists from tackling *Truth and Method*. My discussion in the following chapters will concentrate on key points in that book, hoping to put in clear and straightforward terms the reasons why non-specialists have found it worthwhile in spite of its difficulty. I will also make frequent use of Gadamer's shorter writings on art and culture, particularly those collected by Robert Bernasconi in a volume in English entitled *The Relevance of the Beautiful* (1986c).

To some degree, an understanding of Gadamer requires a grasp of the thinkers who framed questions for him and indicated directions for his thinking, but equally important are his departures from those who taught and inspired him, for he was never simply a follower of others. Gadamer related to even his most influential mentors through a critical dialogue. This critical relationship can be easily overlooked, in part because Gadamer's congenial nature gave him the habit of expressing even very sharp disagreements in respectful, diplomatic, and sometimes even jocular terms.

Gadamer related to even his most influential mentors through a critical dialogue.

Gadamer's life spanned virtually the whole of the twentieth century. Born in Marburg, Germany, in 1900, he spent his youth in Breslau. His father was an accomplished professor of pharmacy but an authoritarian patriarch who had great difficulty condoning his son's attraction to the humanities and the ancient classics (Gadamer 1985, 1–5; Grondin 2003, chs. 2–4). With the appointment of his father to the University of Marburg in 1919, the young Gadamer entered into studies there. The young man's interest in ancient philosophy and literature lead him to work with the great Plato scholar Paul Friedländer. But he came to be utterly captivated by Martin Heidegger's reading of Aristotle, in which Gadamer recognized not only rigorous scholarship but the insight of a first-rate thinker. Gadamer worked in the 1920's as Heidegger's departmental assistant, and Heidegger and Friedländer oversaw Gadamer's habilitation in 1929 (Gadamer 1985, 7–10, 46–50; Grondin 2003, chs. 5–7). But as Heidegger began affiliating himself with the Nazi party in the 1930s, Gadamer reacted with the thought: "he has gone crazy" (1992, 10). Gadamer taught at Leipzig through the Second World War, seeking to preserve the integrity of the university in a time of nightmare, and was appointed rector there when the university reopened after the war in 1946. In his rector's address he denounced the insanity from which Germany was emerging, though privately he had little optimism about how the university might fare under Soviet rule (Gadamer 1985, 93–115; 1992, 15–21; Grondin 2003, chs. 11 and 12). In 1949 he took a position at the University of Heidelberg, the institution with which his name is now most often associated. As his work became known throughout the world he became a frequent traveler, and even after his retirement from Heidelberg he often taught and lectured, including regular courses offered as a visiting professor at Boston College in the United States (Grondin 2003, ch. 16).

While Gadamer generally kept a distance from Heidegger in the later 1930s, his attendance at Heidegger's famous lecture on "The Origin of the Work of Art" coincided with the teaching of his own courses on "Art and History," which are regarded as the laboratory in which the approach to questions of art in *Truth and Method* was first developed (Gadamer 1997, 47). The influence of Heidegger's late thinking on art and poetry can be seen throughout many of Gadamer's most important writings. Though Gadamer had published many papers and had founded a respected journal by the end of the 1950s, it was not until the publication of *Truth and Method* in 1960 that he came to be recognized as an important philosopher in his own right. He did not view that work, however, as a finished project, and he produced a number of further essays on hermeneutics that he brought together as a kind of second volume of *Truth and Method* in his collected works (1986a).

Gadamer is often described as Heidegger's most prominent student. This fact is particularly important to the present book because of the great influence that Heidegger's writings have had on the interpretation of architecture. It was Heidegger, in fact, who had the idea of adapting the hermeneutic circle to the study of human existence. Moreover, it was Heidegger who inspired Gadamer to use hermeneutic thinking to explore the truth-dimension of the work of art (Sharr 2007; Heidegger 1971). Yet there are important points of divergence, as well, between Gadamer and Heidegger, including differences on the subject of architecture. The reader should be aware, then, that when I draw upon insights of scholars such as Karsten Harries, Christian Norberg-Shulz, or Robert Mugerauer (Mugerauer 1994; 1995; 2008), I am using theorists who are influenced most directly by Heidegger, and who connect with Gadamer primarily through Heidegger. As I refer to such theorists I am trying to be careful to select points that have their equivalents in Gadamer, keeping in mind the respects in which Gadamer might have found them following Heidegger a little too closely for his taste.

Heidegger and Gadamer were both closely associated with the "phenomenological" philosophy of the philosopher Edmund Husserl, who was a mentor to both of them. Gadamer appreciated Husserl's attempt to develop a philosophy that overcomes some of the narrowness of natural-scientific approaches to human experience and orients philosophy to the whole of what he called the "life world." For this reason it is fair to associate Gadamer with the "phenomenological" approach to architecture. Here too, however, Gadamer introduces ideas into that movement that are so different from Husserl's as to make it difficult to call Gadamer simply a "phenomenologist." Because Gadamer's principal contribution was a "philosophical hermeneutics," his approach is sometimes called a "hermeneutic phenomenology."

There is hardly a page of writing by Gadamer that does not reflect some influence of the philosophies of Plato and Aristotle, yet he cannot be called a "Platonist" or an "Aristotelian" in any ordinary sense of those labels. From Friedländer, Gadamer learned a style of Plato interpretation that insists that Plato's *Dialogues* cannot be understood if one ignores their dramatic and literary qualities. The Platonic works are not fundamentally a set of doctrines dressed up in a literary expression. They are attempts to capture the process of philosophical inquiry as it naturally unfolds in living dialogue (Grondin 2003, 120–21). What is most important in Gadamer's immersion in the works of Plato is the search for Plato's driving questions and the pursuit of those questions in the spirit of Plato's mentor, the indefatigable questioner Socrates. Plato was certainly interested in formal and mathematical structures of reality, which is what is often singled out as his greatest relevance to architecture, but the Socratic spirit in Plato's thought integrates this interest with an enormous range of questions regarding art, life, morality, and religion, all of which have an equal relevance to architecture. Gadamer sees, in Aristotle's ethics and practical philosophy, a continuation of the Socratic idea that philosophical inquiry is relevant to the problems of practical life, whether they be in the craft professions or in the political life of citizens (1984c 88–138; 1990b, 312–24). The influence of this quality of Aristotle on Gadamer makes Gadamer's thought relevant to the practical side of the architectural enterprise. In sometimes surprising ways, Aristotelian thought can help make sense of the collaborative work of today's practitioners.

> In sometimes surprising ways, Aristotelian thought can help make sense of the collaborative work of today's practitioners.

Gadamer's immersion in the German historical tradition, and particularly in the work of figures such as G. W. F. Hegel, F. W. J. Schelling, and Wilhelm Dilthey, helped him to formulate the questions of the nature of history that are central to his philosophical hermeneutics. Can we say that there are universal truths in human life—e.g., universal standards of justice, truth, and human purpose—when we find so many cultural differences in human history? Hegel had one kind of answer to this question. He believed that concepts develop through history, but that there is a logic to this historical unfolding. But thinkers in the "historicist" school gave less importance to the very idea of universals in history. Perhaps the most widely recognized contribution of Gadamer's thought to the field of philosophy is its effort to steer a unique course among these complex positions on the nature of history and historiography (1990b, 218–31). Finding Hegel's approach to be too speculative and the historicist approach incapable of saying how the past is available to people living in the present, Gadamer's hermeneutics provides unique insight into ways of sorting out the general and particular dimensions of history.

In rejecting Hegel's idea that there is a logic in historical development, Gadamer's views echo those nineteenth-century thinkers such as Kierkegaard and Nietzsche. Kierkegaard insisted that there could be no logic of human existence and Nietzsche argued that thinking with concepts always diminishes the richness of life (Kierkegaard 1992, 118–25; Nietzsche 2006, 114–23, 456–85). One can see such claims as precursors to Gadamer's objections to the formalism of the natural-scientific worldview. But Gadamer's philosophical motivations were also very different from those of Kierkegaard and he regarded Nietzsche's rejection of conceptual thinking as being too radical to permit a productive engagement with the philosophical tradition. While Gadamer thus had less proximity to Nietzschean thought than did his mentor, Heidegger, many Gadamer scholars have shown an interest in making Gadamer's hermeneutics compatible with post-structuralist philosophies that draw significantly from Nietzsche, and especially the deconstructive philosophy of Jacques Derrida. For this reason (and because there has been a good deal of influence of post-structuralist thought on architecture) we will need to consider the relation of certain post-structuralist ideas to those of philosophical hermeneutics.

Plan of the book

In this first chapter I have introduced Gadamer's basic philosophical orientation in a way intended to make understandable my strategy for elaborating his central philosophical ideas and for connecting them with the world of architecture. My second chapter will concentrate on the notion of "play" that is so central to Gadamer's aesthetics, and therefore central to a hermeneutic approach to architecture. A third chapter takes up the best known sections of *Truth and Method*, devoted to hermeneutic awareness and historical understanding, approaching these sections in light of the question of architecture's relationship to its past. Chapter Four examines the question of the relation of hermeneutics to the methods of natural science and their influence on the study of human culture. The issue was the subject of a famous, decades-long exchange between Gadamer and his colleague Jürgen Habermas, but it is also taken up by Vesely, Pérez-Gómez, and others in the context of trying to understand how the technical and artistic functions of architecture became separated from one another in the course of Western history. Chapter

Five explores the relevance of Gadamer's notion of practical philosophy to the more practical dimensions of architects' work. Chapter Six turns to the question that has made Heidegger's philosophy so fascinating for architects, the question of a hermeneutic ontology. Like Heidegger, Gadamer provides a way to think about the meaning of being, with the result that his thought can serve as a means of connecting architecture with that profound question.

In each chapter I shall continue to draw attention to architectural theorists and practitioners who draw upon Gadamer, or who express ideas parallel to Gadamer, or who seem to exemplify points that Gadamer makes. In the brief compass of a book of this sort I cannot begin to give an adequate account of their work, or even to mention all of the authors who have done such work, but my hope is that I have done a sufficient service to both the authors and my readers by drawing attention to a sampling of scholarly efforts. For the inevitable limitations and omissions I offer my regrets and my encouragement to readers to explore beyond what I have offered here.

CHAPTER 2

The Play of Art and the Art of Architecture

The meaning of the German word "*Spiel*" cannot be fully captured in one English equivalent. It can mean "play," as in "child's play," or as in a work of theater. But it can also mean "game," which is certainly something that one plays, but plays according to an explicit structure or set of rules. In the history of German philosophy of art the notion of *Spiel* has long held a central place. Immanuel Kant, most notably, took it to be the quintessential mental activity that defines the experience of art and natural beauty. To engage in play, in the Kantian sense, is to follow connections, relations, and associations among sensations, images, and ideas in a free-flowing manner, yet a manner that has a direction, that seems to be taking one somewhere. This directionality, or teleology, of play does not lead to a factual judgment about the objects of one's play, for that would take the experience out of the aesthetic realm. Rather, the judgment that emerges is a purely aesthetic one: that the object is beautiful—which is simply a judgment about the object's ability to evoke this aesthetically satisfying form of play. Thus the aesthetic judgment, Kant says, is not an objective judgment but a "subjective universal" one (Kant 1952).

Spiel also has a place of importance in German anthropology, particularly in the highly influential 1938 study by Johan Huizinga, *Homo Ludens* (Huizinga 1950). For Huizinga, human culture is fundamentally the product of play. While it can certainly be said that animals play, in human consciousness play emerges in a liberated and liberating form that must be said to be a defining element of human identity and dignity. Huizinga's theories are cultural and historical, but one can see how Gadamer is appropriating them philosophically, combining them with Heidegger's efforts, also undertaken in the 1930s, to argue for the truth of the work of art—i.e., its ability to reveal dimensions of the world and human existence that other kinds of expression cannot equal.

The insistence of thinkers such as Huizinga and Heidegger on the truth-dimension in art and play helps to explain why Gadamer chose to devote the first third of *Truth and Method* to an extended philosophical study of these topics. He introduces the idea of play as a "clue" to the more general nature of hermeneutic understanding. Play is an illuminating form of hermeneutic experience. Its structure reveals a great deal about the general character of the hermeneutic phenomenon. But even as play is thus serving as a clue to the nature of philosophical hermeneutics, Gadamer is simultaneously attempting to strengthen the case for the philosophical *importance* of play—and, by implication, the philosophical importance of art. To do so requires that he recover that importance from certain influential aesthetic theories of the Enlightenment and Romantic periods, which had subjectivized aesthetic truth to the point where art could no longer be seen as bearing the kind of relation to reality that it had been accorded throughout most of human history.

<u>even as play is serving as a clue to the nature of philosophical hermeneutics, Gadamer is simultaneously attempting to strengthen the case for the philosophical *importance* of play.</u>

The problem of the subjectivization of art

The problem can be recognized in an ambiguity in Kant's aesthetics. For Kant, although aesthetic judgments are subjective, they are nonetheless capable of being universal. Moreover, in their universal assertion of beauty, these judgments help to bring together the moral value of the world (i.e., the conviction that human freedom and purpose are at home in this universe) with the sensual experience of the world's natural ways of appearing. Thus art can continue, in Kant's philosophy, to be thought of as prefiguring something important about reality. The directionality that one experiences in the cognitive play evoked by artworks helps to attune the mind to the quality of purpose or teleology in the world (Gadamer 1990b, 54–5). Yet at the same time, the aesthetic judgment of beauty is not in any direct sense a judgment about nature or the universe. It is the expression of the artist's unique ability to call forth cognitive play. When the artist creates something that evokes play and is judged to be beautiful then the art has achieved its purpose, and has done so entirely within the realm of subjectivity. This is the ambiguity of Kant's theory. Indirectly art seems to anticipate something about the world; directly it can only express, and elicit movement within, the sphere of subjectivity.

The aesthetics of the Romantic period, and especially the work of Friedrich Schiller, resolved the ambiguity in favor of subjectivity and individual genius. Schiller saw the freedom afforded by artistic creativity not as a way of relating to the world, but as a means of perfecting the world by moving beyond its prosaic qualities. The purpose of art is to carry the imagination into an autonomous realm of the aesthetic. While this approach clearly adopts Kant's insistence on the importance of the freedom in the play of art, it divorces art from both of the two realities that Kant thought art could bring together: the inner moral life of human consciousness and the outer world of nature. From the very beginning, Schiller takes art to be something superior to reality. The flaw in this approach, according to Gadamer, is to be found not only in its increased subjectivism, but in its very assumption that art and reality are by nature opposed to one another (Gadamer 1990b, 82–3). For Gadamer, art is disclosive of reality, though always in its own distinctive fashion.

For Gadamer, art is disclosive of reality, though always in its own distinctive fashion.

Several theorists have explored the phenomenon of the subjectivization of art specifically in terms of architecture, using historical developments in style and theory to help articulate the situation in which architecture finds itself in the modern and contemporary periods. Dalibor Vesely (to name one who draws explicitly on Gadamer) emphasizes the changes to art's self-understanding that accompanied the emergence of Baroque art, changes that caused an inherent division in architecture's manner of representing. Baroque art emphasizes dramatic and dynamic effects aimed at heightening the emotional response of viewers. In architecture, the result is highly decorated surfaces on both the interior and exterior of buildings. This decorative tendency represents a movement away from architecture as fundamentally an embodiment of symbolic meaning and towards the creation of pleasurable sensations. At the same time, the technologies of architecture cease to be integrated with the work's symbolic purposes, serving instead as the purely technical means of supporting aesthetic artifices (Vesely 2004, 269). Pleasure and technique are both purified, one might say, in this process, but at the cost of being divided from one another. This division sets up the conditions for architectural aesthetics to move along a different path of development from that of architectural engineering, in a trajectory that ultimately finds these two dimensions of architectural creation working, at the beginning of the modern period, at cross-purposes.

Karsten Harries' studies of Baroque architecture yield a similar conclusion. The aesthetic experience that is intended in Baroque styles is one that is self-contained, having no other purpose than a certain mode of imaginative experience. Harries finds theoretical expression of this ideal not only in Kant but in Kant's contemporary Alexander Baumgarten, who identified the aim of art as a perfection of the sensible world, and who characterized the artist that creates this perfect world as a creature possessed of god-like powers (Harries 1997, 16–24). Again, an emphasis on the aesthetic, understood as that which occasions a certain kind of subjective experience, has the effect of separating the artistic dimension from the technological and the functional. Architecture becomes technical construction plus decoration. The paradigm for the building is the decorated shed (1997, 4–5).

For Harries, Vesely, and others who share their interpretive approach, the movement towards subjectivism and aestheticism, towards the goal of "art for art's sake," with its division of the functions and the forms of creation in architecture, has introduced problems for a discipline whose practitioners already faced great challenges to making their art an integral one. To one who embraces the idea that architecture should strive to integrate structure, function, symbol, and artistry the division of architecture's purposes must seem like a surrender of architecture's most definitive calling.

Art and "serious play"

Gadamer's attempt to reverse the trend of subjectivism in aesthetic theory draws, in important respects, on a quality of the philosophical discussions that one finds dramatized in the *Dialogues* of Plato (Gadamer 1980, 70–71). In them one finds Plato's mentor, Socrates, exploring questions, ideas, definitions, and hypotheses not to defeat opponents in debate but to figure out which lines of thought might be the most reasonable ones. This process of Socratic inquiry, which I earlier identified by the term "dialectic," has been called a kind of "serious play." In it there is always a tension between creativity and discovery. Creativity, as every artist knows, is playful. It requires this playfulness in order to be open to possibilities. Truth is serious, often gravely so; but the discovery of truth requires, like artistry, an openness to unsuspected possibilities, so that philosophical inquiry comes to engage playfully with serious matters, yet to take even that playfulness seriously. Recognition of the constant presence of this ironic tension between playfulness and seriousness is essential, in Gadamer's view, to understanding anything about the Platonic corpus. In the serious play of Socrates, consideration of matters such as the nature of justice or of knowledge is liberated from the demands of immediate decision so that the questions can be pursued in a kind of game of disputation among friends, wherein the goal is to let reasons and consequences take the whole of the conversation where they will—which, one may hope, is in the direction of greater truth. This play of Socratic dialectic is emphasized further by Plato's focus on the story of Socrates himself—the fact that Socrates does not disappear behind his ideas but is always living them out (Gadamer 2007, ch. 14; Brogan 2008).

Genuine play in art, for Gadamer, bears this same mark of being capable of seriousness in its very playfulness. In part this is so because freedom is one of the most profound of human potentialities (Gadamer 1986, 130). But by "freedom" here we must not have in mind an endless rambling of curious possibilities. A game orders the potentialities of play into an activity with rules, which, in a certain sense, "plays the players." A work of art transforms play into a specific structure that orders the experience of its audience. A truly successful work of art is stunning in its identity and definition; it hits one like a blow, with a distinctive kind of force; it carries a powerful normative authority, as is expressed in the oft-cited line of Rilke's poetry wherein the archaic torso of Apollo makes its implicit demand: "you must change your life" (Rilke 1982, 61; Gadamer 1986c, 34). The freedom of play indeed has directionalities to it, as Kant rightly observed. Kant sensed that these directionalities are of value because they lead somewhere significant. But by minimizing the parallels between the play of art and the play of truth-seeking, Kant and the Romantics failed to fully appreciate the significance that art has beyond itself.

<u>Genuine play in art, for Gadamer, bears the mark of being capable of seriousness in its very playfulness.</u>

Part of what these thinkers do not grasp is something that Plato, long before them, understood completely: that art occupies a realm that is neither exactly truth nor exactly fiction, a realm that is neither purely known nor purely unknown, that combines the subjective and objective in an undifferentiated way. Plato realized that this realm, which art inhabits, between the known and unknown, is the same uncanny realm occupied by every form of genuine truth-seeking. To claim with Socrates, that in spite of all one may know, one lacks wisdom and lives in ignorance, is to acknowledge that human consciousness inevitably spends most of its time in this state of the in-between. If Platonic philosophy seems, often, to be at odds with the poets and artists of his day, it is not because Plato had no appreciation for art, but because he was in a struggle for this territory of the in-between. In this struggle he had to combat many kinds of opponents, including those who would assert their poetic fantasies as theological verities and those who would insist upon an ancient Greek equivalent of art for art's sake (Gadamer 1980, 43–4; 1986c, 14–15).

The speechless image and the embodied word

It might seem easiest, when attempting to identify the meaning-function in art, to privilege narrative art forms, for these are verbal in the way that factual speaking is verbal. They reproduce the kind of speech found in accounts of things that really happen and the kinds of dialogue in which real people actually engage. In verbal arts one can easily recognize and relate to the meanings that are expressed in artfully embellished language. Similarly, if one is considering a non-verbal art such as painting, one might think that the most accessible kinds of meaning will be found in paintings that depict recognizable people or objects. Such depictions in visual art create a kind of bridge to a narrative. We can see a story being depicted and we can interpret the painting in terms of the story. But there is something deceptive in these paths to interpretation, for the familiarity and easy access may actually obscure the qualities that make the artworks in question distinctly *artistic*. Indeed, because artistic meaning differs so considerably from ordinary ways of meaning-making, an appreciation of the meaning-function of non-verbal images can, ironically, open new ways to interpret verbal arts.

What we are likely to miss if we overemphasize art's proximity to the ordinary is that every artistic representation is a "presentation" and every successful imitation evokes a "recognition." Gadamer uses these terms, "presentation" (*Darstellung*) and "recognition" (*Wiedererkennung*), in technical senses that must be explained. An artistic representation is a presentation in the sense that it shapes, orders, and highlights features of its subject. It presents its subject to us as if to say, "Here, look at it this way." A painted portrait of an important figure, for example, presents the image of the figure in a manner that communicates that importance. We might look at it and say, "What a wonderful likeness!" but what makes it wonderful, from the point of view of art, is not the likeness; it is what is added to the likeness. The work of art is never simply a record of something as it is; it always constitutes an "increase in being" (Gadamer 1990b, 110–18, Vilhauer 36–8).

The work of art is never simply a record of something as it is; it always constitutes an "increase in being."

Art imitates life, then, or nature, or the world, but in a manner designed to help its viewers or hearers recognize something about life, nature, or the world that they might not have grasped before. What is important in this—again, from the point of view of art—is not the success of the imitation as a copy, but its value as a means of recognition. We are speaking here of what has sometimes been called art's power to bring out the universal in the particular or the ideal in the real. But while Gadamer does see some truth to these formulations (inasmuch as presentation always involves a generalizing movement) he cautions against excessive idealism in the characterization of art. One of the most extraordinary features of the work of art is, in fact, that it is able to indicate something general while relinquishing nothing of the importance of its particularity. When the poet, Rilke, describes the archaic torso of Apollo, and concludes with the idea, "you must change your life," these words are trying to indicate what the experience of the work is like; they by no means intend to replace the experience of the work itself. Still less could one turn the meaning of the work of art into a set of concepts that one could take away, leaving the work of art behind. The meaning of the art is something that is retained in the repeated or recollected experience of the work. Works of art share their meaning with extraordinary generosity, yet retain possession of that meaning with a proprietary stubbornness (Gadamer 1986c, 33; 2007, 214–17).

These observations on the nature of art have an architectural illustration in what has come to be known as "Gadamer's floor." Architect Jacques Herzog relates how, when the 96-year-old Gadamer was interviewed in association with a planned exhibit at the Pompidou Center in Paris, and was asked to say what he considered architecture to be, he did not respond in a general way but spoke of a particular experience of his youth.

> In the home of his parents ... there was a wonderful parquet floor in the formal reception room, into which the children (including the young Hans-Georg) were not allowed to enter except on special occasions like Christmas. Describing the piano and billiard table that stood on this bare floor, Gadamer spoke of this surface as something magical—a wonderful wooden floor, immaculately well-kept and polished so that it filled the space with the smell of wax. Every once in a while a friend of his father's would come to visit ... and would, upon entering the forbidden room, always

place his coat and soaking umbrella right down on the magical floor. As a child, Hans-Georg would be horrified that a friend of his father's would do such a thing. He still vividly remembers the image of the polished wooden floor decorated with water droplets from the sodden umbrella.

(Herzog 2001, 115)

The captivating power of this floor emerges, in part, from the extraordinary handwork that fashioned it, in part from the care and ritual that the family brought to it, in part from the fecundity of a child's imagination—yet it is reducible to none of these. The words that describe the experience conceptualize it, thus generalizing its import so as to preserve and communicate it, but they approximate that specific import best by staying close to the experience, expressing it in evocative narrative (or, in the example of Rilke, in poetry) for language in such cases should lead, not away from the experience, but more deeply into it.

Certainly the sense of architecture as a condensation of recognizable meanings in embodied experience is present in Steven Holl's descriptions of his own "archetypal experiences" of architecture. His phrase "archetypal experience" is appropriately paradoxical: the archetypal is sought not in the rarified abstraction of concepts, but in the intensity and singularity of the concrete. One such description draws from his almost-daily visits to the Roman Pantheon, near which he was living as an architecture student in 1970.

> In the tremendous space of the Pantheon, I first felt the passion, the forceful capacity of architecture to engage all the senses. Nearly every morning for months, I passed through the huge double doors and stepped into the spherical silence of this 2,000-year-old work. Each day, its appearance varied with the dramatically changing shaft of light that passed through the open oculus. On rainy mornings, the cylinder of downpouring light contained flashes of raindrops reflected as they slowly fell to the floor and drained into the ingenious marble pavement grooves ... A hazy day rendered the light from the great round orifice more visible, like a solid cylinder of morning sunlight ... The surrounding city fabric, with its various buildings,

> crumbling stone walls, and haphazard, moat-like spaces filled with sleeping cats, stood in astonishing contrast to the pure, hollow interior.
>
> (Holl, et al. 2006, 122)

Few buildings have received more technical and symbolic analysis than the Pantheon, but Holl concentrates here on the complex sensuality that instantiates the meaning and inserts one bodily into it. The communicative power of its intelligibility is not diminished but, on the contrary, increases by this heightened sensual immediacy.

In their project of transforming an enormous power station on the River Thames in London into the Tate Modern, Jacques Herzog and Pierre de Meuron were inspired by the story of Gadamer's floor to think very deliberately about a kind of floor that would assert itself as a unifying architectural element. Though the rough oak plank floor that they chose is far from the handcrafted parquet of Gadamer's youth, it plays a similar role in communicating to gallery visitors a sense of occupying something more than ordinary space. One would be very much mistaken to see such an emphasis on the sensual qualities of particular materials in Herzog and de Meuron, or in Holl, or Juhani Pallasmaa (2009), or Peter Zumthor (Sharr 2007, ch. 5) as a movement away from comprehension and intelligible significance. Such architects should rather be seen as embracing the same enigma that Gadamer finds in every work of art: the mysterious intertwining of endless intelligibility with endless particularity.

This curious tension of the general and the individual is also at work in Gadamer's understanding of the notion of "symbolism." He draws on an ancient source of the word "symbol," which referred to one piece of a broken object that was shared among two people as the indication of a bond between them. When they, or perhaps their descendants, came together long afterwards, the joining of the broken parts would occasion a recognition of the longstanding bond (Gadamer 1986c, 31–4; Tate 2008, 195–9). In an analogous way the work of art is symbolic in that it points beyond itself, but not in a self-effacing way. It is a fragment of the total meaning; it is an essential but incomplete part (1986c, 16, 37, 126). The grasp of symbolic meaning constitutes a realization, as if suddenly coming upon something familiar. The archaic torso is a fragment of an entire ancient world, but when it speaks to one, when it makes its insistent demand, its meaning becomes not only familiar, but immediately and intimately so.

The grasp of symbolic meaning constitutes a realization, as if suddenly coming upon something familiar.

Non-verbal art forms, such as painting, sculpture, instrumental music, and architecture, are "speechless," says Gadamer. But the sense of speechlessness that he has in mind derives from connotations in the German "*vertummt*": not merely silent, but also stammering. The speechless image stammers because it has more to say than it can put into words (1986c, 69, 83). This surplus of meaning in the speechless image reveals another fundamental characteristic of the symbol: that it is multivalent, that it condenses multiple kinds of indication within itself, gathering together different realms of being. Such a function is key to understanding the magical powers that have often been associated with symbols. If a stone carving of a female figure, for example, can carry all at once the materiality of stone, the form of a human, and the signification of divine being, why should it not also combine the powers of all of these within itself? Even if one is not inclined to accept the magical aspect of this power, still one may grant that to see the symbol as a living symbol is to recognize its participation in multiple kinds of being all at once.

While the examples of non-verbal art in *Truth and Method* are drawn from representational, or figurative, styles, Gadamer's later writings make clear that his ideas are equally relevant to non-figurative art, a feature that is important for our present purposes, given that architecture, as an art form, has always been one of the more abstract varieties. One can design a column so that it represents a papyrus stalk or a human figure, but columns that lack such figures can be equally beautiful and significant. Indeed, to modern sensibilities the ornamentation of the former kind of column seems like an unnecessary sculptural addition to the more purely architectural artistry of the unadorned form.

Gadamer insists that abstraction of this sort is not an abandonment of representation (or what the ancient Greeks called "mimesis") but a shift to other kinds of representation (1986c, 24, 36, 103, 128). To simplify imagery down to elemental geometries, flat surfaces, or pure color relationships need not be construed as a departure from the recollection of nature, given that nature is, after all, geometrical, surfaced, and colored. But the abstract artist's movement away from forms that depict objects and scenes can be a way of exploring, highlighting, playing with the qualities that are brought to the fore by abstracting them. This is what the painter Wassily Kandinsky was thinking when he came to the conviction that objects were "harming" his paintings (Kandinsky 1994, 370). There were meanings that could not emerge as long as viewers were lulled into complacency by the familiarity of the things depicted.

Abstraction in this context, so far from minimizing the symbolic import of works of art, can in fact enrich it. Suppression of denotation allows the work to brim over with connotation. By representing no one thing the work can suggest many things at once. In this way the abstract work can enhance the multivalence that is, for Gadamer, definitive of symbols. Even when the ambitions of the work are utterly experimental, there is something that is meant by saying that the experiment was a success, and that is the moment of recognition wherein something is suggested about the ordering of the world. By this potential for recognition the experimental work achieves continuity with the works of the past that inclines us to bestow upon the oddest of things that ancient and venerable term, "art" (1986c, 22–5, 90–92, 128).

Abstraction in this context, so far from minimizing the symbolic import of works of art, can in fact enrich it.

The "decorative" nature of architecture

While Gadamer gives explicit attention to architecture only briefly in *Truth and Method* and other writings, he says enough to make clear that he considers architecture to hold an exemplary place among the arts. His comments, however, can have the opposite effect than the one he intends because he uses a problematic word to describe the quality that gives architecture its unique role: architecture, he says, is "decorative." The word can be misleading, because decoration and ornament are the qualities that Adolf Loos, Le Corbusier, and other modernists attacked as epitomizing superficial aestheticism. Gadamer uses "decorative," however, in an opposite sense— as that which unifies a work of architecture and thus serves as the means of resisting aesthetic superficiality. He draws, here, on Vitruvius' sense of "*decorum*," which refers to the fittingness of a work's form to its meaning and function, exemplified, according to Vitruvius, in the way a temple's structures— and particularly its columns—must be suited to the divinity that the temple is meant to honor (Rykwert 1996, 237–9).

Gadamer adapts Vitruvius' basic idea to his own notion of hermeneutic play. The "fit" or suitability of which Vitruvius speaks has a kind of twentieth-century corollary in Heidegger's account of the way things "gather" a world. To treat a bridge, for example, as an isolated, autonomous entity is to miss not only its architectural function but its very manner of existing within a world of meaning. As a thing it is always caught up in a web of natural and cultural involvements; as an architectural thing it is especially dedicated to the gathering of a landscape, for it draws two banks together, relating a path—and thus a human journey—to the river and its surroundings (Heidegger 1971, 151–8). The bridge orients one within the landscape and complements that landscape by virtue of what it adds. It serves its purpose; it draws significance from its mediation of an environment; yet within the fulfillment of these purposes it realizes its unique artistry. Similarly a building "gathers" by mediating an exterior context and interior spaces. It finds its place within an environment and serves the purposes of its program, with the result that its artistic qualities must both assert themselves creatively and withdraw behind the activities that they make possible and enhance (Gadamer 1990b, 156–8).

There is an analogy, here, to the way a picture frame creates the space of a painting even as it recedes in the experience of the painting. But in the case of architecture the effect is far more profound, for architecture establishes the space in which the whole of an activity, including both the event and its participants, can take place. Moreover, because the function of every art is to enhance the being of things, architecture has this affect on everything within and around it. And because the other arts occur within, or in connection with, architecture, all depend upon the power of architecture to create and shape places. This is what gives architecture its preeminent role. Because it recedes as it serves such purposes, one is inclined to forget its necessary presence—to think of painting, or sculpture, or theater as being autonomous arts—but it is architecture that facilitates their ways of gathering meaning, and architecture is always part of what they will gather.

In this account of the decorative a quality that is often treated as a weakness of architecture emerges as a strength. The artistic dimension in architecture is often characterized as being under threat, and incapable of being a "pure art," because it must always serve its program. While it is of course true that uninspired practical functionalism can undermine architectural artistry, it is also true, in Gadamer's view, that in the most artistically successful architecture the program is entirely integral to the art. It is through program that one becomes involved with architecture in a way that one cannot with most other arts. It is through program that architecture can shape experience by receding into the background. In those surprising moments when an architectural work's artistry suddenly moves into the foreground, much of its power derives from the fact that one is already bodily and intentionally caught up in the nexus of its meanings.

> **It emerges as an artwork only when, in the middle of its use, something wonderful shines forth, as with everything that is beautiful. The experience causes us to pause in the midst of our purposeful doing, for example in a room of a church, or in a stairwell, when suddenly we stand there and remain entranced.**
>
> (Gadamer 2007, 221)

In seeing architecture this way Gadamer is not only extending the Vitruvian sense of *decorum* but is connecting to an ancient use of the term, one that was more or less equivalent to the very idea of "beauty." The beautiful, in this ancient conception, is the form of the good. The attraction of the beautiful intimates the supreme value of the good (Gadamer 1990b, 480–82). By this legacy of meanings, linking the decorative historically to the form of the good, Gadamer has moved the meaning of the term very far from the idea of "mere decoration." He has, in fact, recovered the unity of delight and ultimate value that Kant, with much self-conscious regret, had weakened and that his followers, without much regret, had altogether sundered.

The beautiful, in this ancient conception, is the form of the good. The attraction of the beautiful intimates the supreme value of the good.

The quotation from Gadamer should not pass without mentioning a poignant personal allusion that it contains. As a faculty delegate for the planning of buildings at the University of Heidelberg in the 1950s, he was unsuccessful in advocating for the preservation of a faculty building designed by Freidrich Weinbrenner that had a stairwell, Gadamer said, "so beautiful that I often needed quite a while to climb it, as I would stop every so often" (Gadamer 2006; Rambow and Seifert 2006). One could speak of "Gadamer's stairwell" as an experiential touchstone for the philosopher that bore the same weight and significance as "Gadamer's floor."

Gadamer's highly integral view of architecture caused him to lament the cultural tendency to appreciate architecture primarily in terms of its immediate visual impact. He associated this tendency with the rise of architectural photography. The excellence of photographs in making it easy to share an impression of a building comes at the cost of separating the visual from tactile, auditory, and kinetic imagery. The culture that was promoting a visual sense of architecture was also creating droves of tourists who simply wanted to "see" buildings, as if the whole world were a picture book and travelling was a matter of turning the pages. The sense that the meaning of the building is discovered in its use becomes lost in this process (1986d; 1990b, 87, 156).

On this point one may again notice the consonance between Gadamer's thinking and the work of architects such as Holl and Pallasmaa, who seek to bring all of the senses into play in the interactions that their works evoke. Central to the character of these works are textures and surfaces, the imaginative sense of the weight of forms and materials, the sense of movement and discovery that unfolds as a person moves through rooms, or as light falls on surfaces through the course of the day and the changes of the seasons. Such qualities bring dimensions of depth and time to the work, for they are realized gradually as one gains experience with the work's manner of being. Gadamer's belief in this kind of gradual emergence of architectural meaning made him reluctant to pronounce upon the success or failure of particular buildings. One would have to spend time with a building, he thought, to discover everything that it was attempting to accomplish. But when the integral artistry of architecture steps to the fore, then everything that happens within it may seem to resonate with its power. Great structures bring their artistry to the whole of a life.

Questions of balance

Between the self-assertion and the self-effacing forces within a work of architecture there must be a balance, and it is to this balance that every architect properly aspires. A design can be bold and striking, it can be elaborately ornamented, and yet it must also have a strategy for receding in relation to that which it seeks to facilitate. Architecture that is just getting in the way, no matter how artfully, is not functioning as architecture. The problem can be considered in light of certain ironic–historicist postmodern styles of the 1980s. Do the extravagant and quirky historical references of James Stirling's *Neue Staatsgalerie* overwhelm the experience of the overall work with heavily visual references that have not been integrated? Is the very point to exaggerate the aesthetic to the point of making a statement rather than an integral work? Gadamer was perhaps inclined to think so, but would not judge without a solid familiarity with the building (Gadamer 1986d). It is fair to say, however, that at some point an experiment such as this, with its striking playfulness and exaggerated references, loses its balance. When architects say "it is not serious" they do not mean to denigrate play as such, but to resist the return of an aestheticism that does not appreciate the great things to which play can be dedicated.

When architects say "it is not serious" they do not mean to denigrate play as such, but to resist the return of an aestheticism that does not appreciate the great things to which play can be dedicated.

It is hard to imagine a more playful architect than Frank Gehry, yet his best works are taken very seriously. Leaving aside, for the moment, the question as to how philosophical deconstructivism might be relevant to the directions that his work has taken, we can say that he has taken sculptural possibilities that were explored throughout the modern period and has furthered them through his own inventiveness and the introduction of computer-aided design and manufacture. The sophisticated design process has brought the finished work closer to the humble workings of the artist's hand. Since the beginnings of his experiments in the 1970s Gehry has been dogged with charges that his approach is sculptural *rather than* architectural or that the aesthetic forms of the buildings are mere additions rather than integral components. But it may well be that it is these readings, rather than the buildings, that are superficial. There may be unique and interesting ways in which Gehry is achieving the sought-for balance.

Some brief observations may serve to open up these possibilities. Gehry's buildings, I would say, are thoroughly architectural in the sense that meticulous attention is paid throughout the creative process to the program and its arrangement. The result is a building that one can navigate, however unusual its shapes and contrasts. One's intuitive motions through interior space have been anticipated. This accommodation of the body-subject in the Gehry building establishes a relationship to the sculptural forms that have resulted from a body-centered design process. To combine bodily elements in this way, I would propose, constitutes a kind of integration. It is not the obvious integration represented, say, by concrete build-through, but it is a complex imaginative integration attuned to qualities of human embodiment. It seeks a decorative balance, in Gadamer's sense—alternately asserting form and deferring to the sensibilities of the building's users.

If Gehry's spaces are surprisingly navigable they are also surprisingly recognizable. As one walks around the lobby of the Disney Concert Hall in Los Angeles, for example, one can recognize the spaces of a concert hall, even where they differ visually from any concert hall one has seen before. The exterior of the building invites many imaginative associations, as in the "full sail" sweep of the entry façade, but among them are invitingly recognizable architectural associations. One finds courtyards and gardens; one can wander along the top of the exterior walls as if visiting a European castle; one can look up and down from all sorts of fascinating vantage points. In the Stata Center at the Massachusetts Institute of Technology one can recognize elements of a small villa, with public spaces such as a miniature amphitheater. The forms of the complex are certainly playful, even comical in some ways, but they maintain an imaginative connection to architectural traditions that have strong resonances.

Play as clue and instance

In the course of freeing the phenomenon of play from the limitations of aestheticism, Part One of *Truth and Method* allows the structure of play to serve as a clue to the structure of the hermeneutic pattern. What should by now be evident is that to call it a "clue" is truly an understatement. The play of art constitutes full-fledged experience of the hermeneutic phenomenon. While the purposes of art are different from those of scholarship, or history, or philosophy, the specifically hermeneutic dimension of art is not. Moreover, because art is embedded in culture and history, the hermeneutic of art does not just parallel these—it is continuous with them. Works of architecture, in particular, Gadamer says, "do not stand motionless on the shore of the stream of history but are born along by it" (1990b, 156–7). The hermeneutic reciprocity of architecture and history can reach a point, in fact, where the separation of the two seems arbitrary.

Frank Gehry, Disney Concert Hall, Los Angeles

Works of architecture, Gadamer says, "do not stand motionless on the shore of the stream of history but are born along by it."

Disney Concert Hall   Disney Concert Hall

Frank Gehry, Ray and Maria Stata Center, MIT

For this reason, as Part Two of *Truth and Method* moves to questions of the nature of history and the interpretation of historical texts, it does not fundamentally leave behind questions of art and architecture. Architectural theorists who are influenced by Gadamer find Part Two to be as relevant to their work as Part One. Their reasons for holding this view will become evident as we move more deeply into the elements of Gadamer's hermeneutic of historical understanding.

CHAPTER 3

Historical Understanding and Architecture's Past

Given the complexity of the working of the hermeneutic pattern in the context of human history, it may be helpful to introduce the ideas with an extended analogy. Imagine a woman who has lived all of her life in a region of villages bordered by mountains. The peaks of the mountains define the region with their jagged horizon line. Beyond the mountains, she has been told, there is a different world, a place where people speak another language and where great numbers of them live in a large city. She has heard reports of the life in this city, but they all conflict. The people that she knows who have visited the city seem to have differing interests and perceptions. Some say it is a wonderful place, some find it frightening, but everyone agrees that it is a completely different sort of settlement where people live lives that seem nearly incomprehensible to an outsider. Then one day she decides to make the journey to see for herself. She travels along the road to the mountain pass. She stands at the place that had been the limit of her vision for the whole of her life and she looks, for the first time, beyond the mountains. In this moment we may ask, what has she brought to this place, the place that was a point on her horizon but now opens upon another world? We must say that she brings everything that she has become by virtue of living within the world delimited by that horizon—her language, her knowledge, her habits and sensibilities. They are hers in the sense that they define her identity and character; but, of course, they have not *come from* her so much as from the long history of the people of the region of villages. Even now, as the horizon ceases to be a horizon and becomes the point of entry to the world beyond, she remains the person that was thoroughly shaped by the community enclosed by the mountains, and she carries the world of her community with her.

The idea of horizon

The literal horizon line of the mountain peaks in this tale serves as a metaphor for Gadamer's conception of the cultural and historical horizons within which every individual and community lives. On the one hand, every such horizon functions as a limit. It delineates a set of beliefs, stories, ideas, customs, shared experiences, and dispositions that make a people who they are. Beyond the horizon are the things that they do not know and cannot imagine, including things that they cannot even ask about because they do not know how to ask, or why one would ever want to ask. The world of this horizon will never cease to be one's origin, one's home. Within its compass are feelings, habits, sensibilities, and associations that one may never be able to entirely alter. They mark one with the stamp of a culture's history (Gadamer 1990b, 302–4).

Yet the horizon is also the means of being open, for the culture that one assimilates from one's youth is what makes the world familiar and life livable. Like the mountain pass, the horizon itself forms the only means of venturing beyond the horizon. Like the physical horizon, the cultural horizon functions both as a limitation and as an opening to everything that transcends it.

Like the physical horizon, the cultural horizon functions both as a limitation and as an opening to everything that transcends it.

Architects and designers know just how deep a horizon goes in the human psyche. They see it embedded in language, but equally in gestures and bodily habits—the way people move within public spaces and congregate together, the way they orchestrate their lives amid public and private spaces. The horizon is manifest in all of the feelings and images that go along with these dimensions of experience and activity. It determines which places feel comfortable and home-like and which ones require a difficult accommodation. If architecture relates to culture as the book relates to literacy, then the architect must be aware of cultural horizons and be able to mediate them.

But let us follow our tale further. The woman from the village descends from the mountain pass and makes her way to the city. At first everything is incomprehensibly strange, but she finds people who can help her. They know something of her language and can translate for her. They anticipate her needs and show her ways to take care of them. Some of them take advantage of her naïveté, but she learns to recognize genuine hospitality and she appreciates it deeply when the city dwellers offer it. They have projects and cares that she never had, but she finds these things more intriguing than not. She wants to get to know this way of life and she decides to remain in the city. Over time, she learns how people manage to live with one another in this crowded and bustling place. New-found friends help her to see which of her habits mark her as an outsider. Although the din of the city is always in her ears and at times she does not think that she can stand it any longer, still the list of ways that she prefers the city to the village begins to lengthen. One day, she finds that she is no longer translating the words that she hears into her native tongue and she speaks the language of the city fluently. Soon after, she is dreaming in it.

When one encounters another horizon—whether it be that of a different culture, or a different historical age, or even a different individual—the experience will be shaped by one's own horizon, both in ways that one can recognize and ways that one cannot. Because one will do things and think things in this situation that are oblivious to the other horizon, we say that one is "prejudiced," that one is acting on assumptions from one's own horizon that may not grasp at all the realities of the new encounter. "Prejudice" is rightly used as a pejorative term to describe an overconfidence about one's assumptions that closes a person to qualities that are different yet equally worthy in others. The word rightly stigmatizes people who confine their lives to stubbornly narrow horizons. But often this pejorative meaning is used too broadly, to condemn the very fact of having a horizon. A horizon is indeed the total set of one's prejudices, but it is also the threshold for all that exists beyond the horizon. The movement beyond one's prejudices must begin with those prejudices. One should recognize them as the starting point for the encounter with what is new and unfamiliar, and one should alter or abandon those that end up distorting the things that one experiences. If one remains open, then new experiences, new discoveries, new moments of understanding will gradually replace the assumptions with which one began. But a number of one's assumptions will ultimately turn out to have been more right than wrong. A prejudice is not false simply by virtue of being a pre-judgment; some pre-judgments turn out to be true. The pejorative sense of "prejudice" should apply, then, to people who are unwilling to put their prejudices at stake, who fail to put their horizon at risk in the interaction with another horizon. It should not apply to every prejudice as such (Gadamer 1990b, 277–9).

This view contrasts with what Gadamer calls the Enlightenment's "prejudice against prejudice" (1990b, 271–3). The rhetoric of the scientific revolution sought to cast everything traditional in a negative light. Writers such as Francis Bacon and René Descartes imagined the whole body of human knowledge as a great, extravagant edifice built upon a weak foundation. They considered it pointless to try to do anything with the building. It had to be torn down and rebuilt upon new foundations. Notice how this metaphor implies a wholesale rejection of tradition, taking all of tradition as naïve and prejudiced, therefore unreliable. They do not want the new foundation to be another tradition, but rather to be a scientific method by which one can prove *for oneself* what is true and what is not (Bacon 1960; Descartes 1993). But in Gadamer's view it is the Enlightenment thinkers who are displaying their naïveté in these matters. To think that one can magically step out of one's horizon by the mere application of a method is to fall far short of grasping the reality of horizons. Living in the resulting imaginary state that is supposedly free of prejudice, one is far more likely to be oblivious to the ways in which prejudices are actually controlling one's interpretations of experience. The very idea of a methodical freedom from prejudice functions, in other words, as a supremely distorting kind of prejudice.

To think that one can magically step out of one's horizon by the mere application of a method is to fall far short of grasping the reality of horizons.

Gadamer's view contrasts, for similar reasons, with some popular notions of toleration and multiculturalism. One might think that simply having an attitude that tolerates and even celebrates differences could overcome the problem of prejudice. But how can one know that the differences are tolerable if one does not understand them? Is there not a risk here of falling into a kind of sentimental relativism that actually makes little effort to move beyond an initial horizon? For Gadamer, the adoption of a set of attitudes is insufficient to address the problem. One ends up either abandoning one's critical faculties or failing to encounter the genuine difference of the other perspective, or both. There is, in his view, no substitute for engaging in the concrete process of building genuine intercultural relationships and shared experiences.

In this light it should be clear that critics of Gadamer who accuse him of having merely endorsed the inevitability and value of prejudice have missed the point. When it comes to overcoming prejudice, Gadamer is in fact raising the bar, not lowering it, by insisting that only a persistent and particularized effort can make headway against distorting prejudices. Key to the process is becoming aware that one inevitably inquires from *within* a horizon, and that the nature of this horizon and the possibility of altering it can be fully realized only in the course of engaging with another horizon.

But how does this process happen? Primarily it occurs through the cumulative working of the hermeneutical pattern. As the woman in our tale begins learning the ways of the city, her first attempts are born of her native assumptions. Where she sees those assumptions failing to make sense of things, she may begin to put her horizon at risk. She may begin to form an understanding of the problems to which the customs of this new place are solutions. She will have moments of recognition: "In the village we did it that way, but in the city they do it this way." She sees that she does not have to learn how to live all over again; she can make a kind of translation. In the process she will not abandon her critical faculties. She will recognize, "In the village the sly ones deceived you that way, but in the city they do it this way."

In the ideal case there occurs what Gadamer calls a *Horizontverschmelzung*, which is normally translated as the "fusion of horizons." It is something similar to the experience that one undergoes when one can finally drop the whole apparatus by which one translated a new language into one's native tongue and one simply thinks and speaks in the new language. When this happens with regard to a whole horizon—i.e. that one can simply make one's life within it—that is when one has truly achieved an "understanding." Understanding, at this level, is not simply cognitive or verbal; it is existential. It alters the very foundations of one's approach to life (Gadamer 1990b, 305–7).

Again, such a transformation cannot happen in the abstract or by simply cultivating feelings of openness and sympathy. The change comes about in the process of developing relationships and interacting in the places where different horizons hold sway. The word "fusion" captures some of the sense of what occurs in the transformation, but the English word is limited if it suggests that the two horizons are simply bonded together. What in fact happens, to the extent that there is success, is that a third reality emerges, something that is born of the two horizons but is equally the product of the new experiences and relationships that have formed. The fusion is never complete, for there is no end to the discoveries that one can make about another individual or another culture, no end to the experiences of hermeneutical recognition that one could have. But the process is limited, *de facto*, by the limits on our time and abilities, by the finite scope and span of our lifetimes.

The interpretation of texts

When one opens a book one opens one's mind to what it might have to say. Yet one also brings assumptions as to the sort of things it *could* say, and these assumptions inevitably form the working interpretation with which one begins. Thus in reading texts, a horizon sets an initial limit to the possibilities for interpretation even as it grants one access to the work. The writer of the work was also informed by a horizon, of course. The writer was attempting to articulate something—a story, say, or a body of thinking. There were matters of concern with which he or she was wrestling, and through activities of "serious play" the author achieved some sort of articulation, expressed in terms of the author's horizon and perhaps the attempt to move beyond it. In this circumstance, as in the others that I have described, "understanding" the author cannot simply mean construing the work in terms of one's own horizon. It must aim at some sort of fusion of horizons.

"Understanding" the author cannot simply mean construing the work in terms of one's own horizon. It must aim at some sort of fusion of horizons.

The woman who is new to the city, once she begins mastering the language of its inhabitants, can test her assumptions in dialogue with the city dwellers. They actively help her along the way to a new understanding. In reading a text, however, one does not have the author available for this kind of dialogue. The effort to understand must therefore substitute, in some manner, for the lack of interaction. For Gadamer, the most reliable way to do this is to look in the text for the patterns of inquiry or exploration that are carrying the author forward in his or her undertaking. By sharing, through the text, in the matters of the author's concern, one may come to be moved by them as the author was moved. Here one is not trying to enter into the psyche of the author; one is engaged fully with what is written; but one is building a relationship with the subject matter of the text that parallels the relationship that the author had to it. In this process one is trying to recognize the questions to which the author is responding, or the experiences that the author is trying to express. One is engaging in something which, if it is not exactly a dialogue with the author, brings into play some of the key elements of a productive dialogue (Gadamer 1990b, 362–79).

Gadamer's approach to texts, then, differs from theories of interpretation that would locate the authority for the interpretation in the intentions of the author. Certainly one can speculate about an author's intentions, and such speculation can be a useful part of the interpretive effort. But because the author is not present, because the assertion of intentions *is* speculative, interpretive authority must rest elsewhere. It must rest in the meaning that is contained in the text, that reflects the efforts of the author, but also the world with which the author was engaged and the horizon that structured that engagement. The authority of the interpretation derives from the interpreter's hermeneutic involvement with all of these dimensions. Nor is this necessarily the case only when the author is absent. The poet Galway Kinnell, for example, in a question session after he had once done a reading, was asked by a member of the audience to explain a single line in one of the poet's works that had always puzzled this person. "I've been waiting for years," the audience member said, "to ask you exactly what that line means." But Kinnell disappointed him. "I don't think that I am actually the best person to try to say what that line means," he responded, "because my interpretation would be so colored by what I *wanted* it to mean. I would be very biased." Kinnell here denies that his intentions are the best measure of meaning, pointing instead to the difficulty of the struggle to capture meaning in verse. In so doing, he invites the questioner into a reciprocal involvement with that struggle rather than releasing the questioner from the effort with a simple answer.

If Gadamer resists a subjectivist approach to the meaning of texts by limiting the authority of the author's intentions, so too he opposes an objectivist approach that would treat the author primarily as a product of his or her time. According to this ostensibly empirical approach, we can understand an author after the fact better than the author did himself or herself, because the tools of historical research provide us with a knowledge of cultural trends and influences that would have affected the author without the author being aware of the fact. While there is surely some truth in this notion of the value of hindsight, from Gadamer's point of view it overlooks several key parts of the hermeneutic problem. For one, while we cannot interpret a text by entering the subjectivity of the author, we must not forget that the text *is* a product of a human subject who exercised freedom along with channeling the forces of history. For another, the objectivist approach imagines history as a march of progress, so that "we know more now." But in fact what occurs with the movement of history are shifts in horizon, so that we never simply know more of the same (Gadamer 1990b, 192–3, 204–12). Rather, we approach matters differently, with different questions, assumptions, expectations, frameworks, and purposes. Hence a hermeneutic that illuminates these shifts in horizon is needed. Furthermore, the passage of time inevitably brings the loss of knowledge along with gains. We may no longer have the kind of background that is needed to make sense of a text from another age. We will understand the text differently than the author did, in some ways better, perhaps, and in some ways worse. But we will fall into all sorts of interpretive traps if we do not grasp the role that horizons play in the process.

The goal of Gadamer's hermeneutics, then, is not to return to the original world of the text nor to the mindset of its author, and the goal is not to assess the author from a superior vantage point. It is, in the first place, to learn how to let the text speak again—not within its own horizon, but in communion with the horizon of we who are living and are attempting to make sense of it. In this respect Gadamer draws heavily on the treatment of the problem of "application" in the history of the development of hermeneutical theory (1990b, 307–11). The framing of the hermeneutic problem originally took place in fields such as law and biblical scholarship where the point was not merely to achieve a conceptual grasp of a text's meaning, but to know how to put it to work in a living community. It is this quality of application that Gadamer is extending to the interpretation of every text. Every interpretation involves a dimension of application because every interpretation takes place within a living horizon.

Horizon and history

An irony in the study of history, from the point of view of philosophical hermeneutics, is that horizons pose limits to our access to the past, yet at the same time they are products of the past. In this irony, however, there lies a clue to the unraveling of the hermeneutic puzzle as it emerges in the study of history. As we have seen, in Gadamerian hermeneutics the approach to another horizon requires bringing to light qualities of one's own horizon. But in the historical context, because the present horizon reflects the influence of the past, the appropriation of one's horizon is in part identical with the discovery of the other horizon. One may encounter the past, in other words, through its ways of shaping one's present experience of the world. Gadamer's term for this relation of the past to the present is *Wirkungsgeschichte*, a word that has posed difficulties for translators. It is "effective history," or "history of effects," or the "influence of history," or the "working of history" (Gadamer 1990b, 300–302). Each of these captures something of the German meaning, though none of them does so completely and none sounds terribly natural in English.

An irony in the study of history, from the point of view of philosophical hermeneutics, is that horizons pose limits to our access to the past, yet at the same time they are products of the past.

> We can imagine the woman in our tale coming to an awareness of the effects of history. As she begins learning the language of the city-dwellers, she notices that some of the words are the same as those in her native language, and some reflect only minor variations. She begins to realize, furthermore, that the shared words tend to refer to things that are made in the city, or came along trade routes through the city. She begins to see that the history of the relationship of the city to the villages is sedimented, in many ways, in her own language. Her awareness of this fact now gives her a new way of looking within the resources of her own horizon for ways to recognize meaning in the world of the city. The experience gives her an ongoing question: What else in my life in the village bears the mark of the city? How else has the city been influencing my life all along?

Once one has grasped Gadamer's notion of horizon, this further insight into the role of history in forming horizons is fairly straightforward, and to give simple examples of it, as I have done, is relatively easy. But to fully incorporate it into one's thinking—to develop, in other words, a habitual *Wirkungsgeschichte Bewusstsein*, or "consciousness of effective history"—is a far more difficult affair. It involves developing an awareness of the historical dimension in all of one's thinking and acting (1990b, 301–2). Were this awareness to become common in the scholarly study of human society, scholarship would look very different from its present-day forms. One can imagine researchers trying to bring their assumptions to the surface of their studies rather than hiding them behind the rhetoric of expertise and scientific detachment. One could see them mining those assumptions for insights into the ways that history has molded their own intellectual horizon. Historians, in particular, might be much more explicit in exploring how the assumptions of the present pose obstacles and opportunities for making sense of the past.

Interpreting the past, for philosophical hermeneutics, is always a matter of retrieval. One recovers something of the past in the present, drawing upon both the past and the present. History does not give us truths that remain uniform across all of time. It does not even give us a single set of questions. Every thought and every question is formulated in some means of expression, and each of these bears the indelible mark of a horizon. When truths transcend horizons it is because they are recognized anew in new horizons. They are recognized in a manner analogous to the recognition of familiar meaning in the experience of a work of art. In making these claims Gadamer is drawing on Heidegger's philosophy of truth. Heidegger often names truth with the ancient Greek word *aletheia* with its suggestion of an uncovering or disclosure that is never complete, but always brings with it the "closure" that surrounds it (Gadamer 1994, 63–4). For Gadamer, too, historical truth is always limited, and the passage of time always brings with it the loss of avenues into the past. Horizons open, but they also close.

It is fair to ask, as many commentators have, whether this Gadamerian view of historical truth does not imply its own kind of historical relativism—i.e., the implication that truth cannot transcend the particular historical culture in which it has come to light . To a number of commentators it has seemed obvious that Gadamer falls prey to such a relativism. Were they correct there would be an irony in the fact, because Gadamer offers his hermeneutics expressly as a means of overcoming the "entanglement in historicism" that plagued historiographers such as Wilhelm Dilthey and Ernst Troeltsch. Gadamer regards these thinkers as laudable for understanding the pervasiveness and depth of the role of horizons, but he faults them for being unable see how horizons are also forms of openness. The openness of horizons—limited, to be sure—makes room for the perpetuation of truth through continual rediscovery and reformulation. In it there is always a mixture of sameness and difference, but in a combination that allows the truth to "speak again."

The openness of horizons makes room for the perpetuation of truth through continual rediscovery and reformulation.

An analogy is possible here (though it is by no means a perfect one) to the case of Albert Einstein. Einstein's principal contribution to theoretical physics was a theory of relativity, and yet no one accuses Einstein of being a "relativist" in the popular sense of the term. Why not? Because although he denied the existence of a single universal frame of reference for all motions in the universe, he was nevertheless able to identify rules governing the transposition of physical laws across multiple inertial reference frames. By analogy, one could say that hermeneutics is the activity of transposition across horizons. But the complexity of the transposition prevents the prescription of any single set of rules; the means of transposition must be discovered in the effort itself.

There is another (and more apt) analogy that one can make to a characteristic of Socratic discourse. Socrates always enters a new philosophical conversation with a remarkable freshness and childlike enthusiasm. The discussion may concern matters about which Socrates made up his mind long ago. But he always welcomes the opportunity to be instructed and to be refuted. In many cases these scenes have an air of comedy to them, because the reader can immediately see that Socrates' dialogue partner is very confused and has nothing to teach. But one should not thereby doubt Socrates' sincerity. A truth, for him, is not an opinion that, once reached, must never be questioned again. On the contrary, a truth remains true by being put to the test over and over again, in the face of new insights and in conversation with people who have very different points of view. Socrates' conviction that truth only lives in dialogue gives him an intuitive sense of the need always to be struggling with the reality of multiple horizons.

Again, Gadamer is not embracing relativism when he observes that our grasp of truth is only partial; this is simply to make the obvious point that as human beings we are finite. There is a unity and constancy to truths inasmuch as they are recognized as the same in many different contexts, but there would seem to be an indefinite variability to those contexts. This means that there is always a sense in which truth cannot be mastered. This statement may seem pessimistic, and yet there is something we love in this infinity of possibility. It is what is so prized about a work that we call a "classic" (Gadamer 1990b, 287). One can return to a classic over and over again to find further meanings that one had never suspected were there. Its truth seems to blossom into endless varieties of implication.

Hermeneutics and history in architectural theory

Every architectural theorist must wrestle with the question as to the relation of architecture to its past, a question made all the more problematic by virtue of the radical break that modern architecture sought to make with that past. Gadamer is relevant to the work of a number of these theorists, though they do not necessarily draw upon him in overt or deliberate ways.

Karsten Harries takes as a point of departure Sigfried Giedion's claim that the task facing architecture is "the interpretation of a way of life valid for our period" (Harries 1997, 2–4). The claim obviously raises two further questions: "How does architecture interpret?" and "What way of life is valid for our period?" Harries is drawn to the idea that architecture interprets by embodying a cultural ethos—the character or spirit that pervades the activities of a community. This he refers to as the "ethical function" of architecture. Harries sees great relevance in Heidegger's existential hermeneutics, because while it is far too original and polemical to be called traditional, still it is fully immersed in concerns that have echoed throughout the whole of philosophy's history. In Heidegger one finds the potential for new ways to articulate the depth of meaning in architecture, using poetic language that speaks to the ultimate concerns of human life on Earth. Heidegger's formulations recall traditional questions of ultimate concern, but without repeating the religious beliefs and metaphysical assumptions of the past. In this respect, then, Heidegger reveals ways in which architecture can be interpretive for our present time.

Harries is drawn to the idea that architecture interprets by embodying a cultural ethos—the character or spirit that pervades the activities of a community.

There are limitations, however, in the directions that Heidegger indicates for architecture. Heidegger's thinking always has its own philosophical goals in view, such that his aim in addressing the meaning of architecture is less to present a philosophy of architecture than to draw architecture's ways of meaning into his larger philosophical project of thinking the meaning of being. Moreover, the examples that Heidegger uses—the Greek temple and the Black Forest farm house—do not speak to the contemporary issues to which Harries is addressed. In fact, they suggest a kind of romanticism that is merely a reaction against modernism and is, in light of its echoes in Nazi ideology, alarming (Harries 1997, 157–66).

The work of the Norwegian architectural historian and theorist Christian Norberg-Schulz (1926–2000) has played an important role in broadening and applying Heidegger's brief forays into the question of architecture. In books such as *Genius Loci* and *The Concept of Dwelling* Norberg-Shulz begins with Heidegger's articulation of the way that architectural creations such as the bridge gather the meaning-dimension of a landscape, but he then uses such examples as clues to asking the question regarding the spirit of a place in any historical context. From a Heideggerian starting point Norberg-Shulz goes on to explore symbolic types such as path and place, earth and sky, the orientation of a settlement within a landscape and the complementary additions that a settlement makes to that landscape. This typology of meaningful forms is aimed at legitimating the meaning function in all building, returning the sense of "dwelling on Earth" to every dimension of design and construction. The approach is characterized also as recovering the "figurative" aspect of architecture. The figures and shapes of traditional architecture have powerful connections to human embodiment in environments, and it is no wonder, therefore, that the public has found difficulty in relating to the types of twentieth-century architecture that have abandoned those figures (Norberg-Schulz, 1979; 1985).

For David Leatherbarrow, however, to define the continuity of the present with the past in terms of figural elements, as Norberg-Schulz does, reduces architecture to style, losing sight of the uniquely varied but integral set of challenges that architecture presents to its practitioners (Leatherbarrow 1993, 80–81). No less interested than Norberg-Schulz in the human connection to architecture, and no less immersed in architecture's history, Leatherbarrow wants to find continuity not in the particular forms that solutions take, but in the persistence of the problems that architecture must solve to suit human needs and desires. The same set of problems can yield very different-looking solutions, yet there is a degree of equivalence in the results. To interpret a building in relation to its time requires seeking out the questions to which it is a response. Examples of these questions would include the question as to how to site the building, how to imagine a suitable enclosure, and how to realize the potentialities of materials. Leatherbarrow offers a vivid example in the famous building by Adolf Loos (the "Looshaus") in Vienna's Michaelerplatz. The outrage that it provoked by preferring clean lines to ornamentation and by displaying the inherent beauty of its materials rather than shaping them into decorations was based on an equation of architectural character with visual style (1993, 58). It missed the important ways in which Loos's building

Adolf Loos, "Looshaus," Michaelerplatz, Vienna

represents a compatible kind of solution to the same problems that the older buildings around it address.

To interpret a building in relation to its time requires seeking out the questions to which it is a response.

In light of these claims Leatherbarrow could be characterized as taking a more Gadamerian than Heideggerian approach to the question regarding architecture's relation to its past and its present. In a manner that recalls Gadamer's openness to modern and experimental art, Leatherbarrow is insisting that the recognition that allows one to understand and identify with a work of architecture may not be obvious. It may require some involvement with the work before the connection can be made. But in this activity of interaction with the work, one may come to appreciate not only the legacy upon which it draws, but also the inventiveness that grants it a unique character, and in this involvement with the sources of the work's creativity there can emerge a profound and personal connection.

Leatherbarrow's book *Uncommon Ground* examines the historical development of the topographical idea of horizon in architectural design with results that are intriguing in light of Gadamer's philosophical use of that term. Leatherbarrow observes, for example, that an attunement to the role of the horizon can help the architect resist a confining over-design of spaces. A horizon anchors a site and contributes to the definition of a place, but it also draws one beyond the immediate. Part of its significance is in the way it recedes, and in so doing breaks up the restrictive coherence of an overdetermined environment (2000, 159, 173). Again, Leatherbarrow's interest in the history of the role of horizon in design is part of the effort to think how modern design practices can carry forward a vital element within the history of architecture. In a statement that has a particularly Gadamerian ring to it, he says, "Tradition can represent either the dead faith in what is now past, or a living, reanimated faith in what has been shown to be still relevant" (275). Adrian Snodgrass and Richard Coyne draw even more openly on Gadamer in making this point. Gadamer provides, for them, an explanation for the historical thinking that is inevitably involved in design activity, but he provides a way of affirming and cultivating that historical thinking that does not merely recall the past but orients its insights to the present and the future (Snodgrass and Coyne 2006, 131–46).

The hermeneutics of sacred architecture

Sacred architecture has always been a central focus of art historians because sacred buildings have a special permission to symbolize the place of humans in the cosmos and their orientation to the transcendent beyond. Sacred architecture functions as one of the most comprehensive forms in which a culture articulates itself through its built environment. In recent years

Gadamer's work has been influential in the ongoing development of the phenomenological approach to the interpretation of sacred architecture. The art historian has the challenge of treating the works that he or she studies both as historical artifacts and art. To experience the works *as* art requires drawing upon one's own horizon; to understand the works in terms of their originating horizon, however, requires an openness to different, and perhaps quite alien, horizons. The problem puts the art historian in the classic situation of hermeneutics: how to respect the historical differences of horizons while bringing out the works' continued power and relevance?

Some basic implications of Gadamer's thought for the problem should be clear. At the very least, the problem is not to be solved by taking a position outside of the historical movement of cultural self-interpretation, as is done by many who develop classificatory schemes of symbolic forms or archetypes. While such a strategy may be quite legitimate in its identification of patterns, similarities, and historical trends, the classificatory scheme can obscure as much as it illuminates by simply fitting the meanings of other horizons into the categories of one's own. The sacred symbols of the past cannot speak to one, cannot really function as symbols, once they have been pinned, like butterflies, to a specimen board. Yet the Romantic alternative of trying to imagine how the symbols would have been experienced by those who made them is subject to Gadamer's criticism that this project cannot help but be a largely fabricated construction—and again, one that is fashioned out of the resources of one's own horizon rather than through an encounter with another horizon.

The sacred symbols of the past cannot speak to one, cannot really function as symbols, once they have been pinned, like butterflies, to a specimen board.

Hence one finds in the work of Lindsay Jones and Thomas Barrie the attempt to steer a hermeneutic "middle ground," one that draws on formal historical analysis, but does so with Gadamerian expectations in place—above all, the expectation that understanding is something that cannot be brought *to* the inquiry but must emerge *from* it, the expectation that the meaning of symbols will emerge in a dialogical orientation that is willing to bring one's own horizon into view and to let it be challenged by the horizons that one studies (Jones 2000; Barrie 2010).

Architecture's past in present design

Even the most decidedly modernist architect aims at some kind of continuity with the past. The pure geometries of the International Style, for example, can be read as distillations of classic forms. The monumentality that the ancient architects could only dream of is realized in spectacular fashion by the modern skyscraper. The democracy symbolized in the architecture of the forums and law courts of ancient Greece is embodied more literally in the modular designs that distribute dwelling spaces across a wide swath of today's population. While such historically minded intentions are not hard to find, the measure of their success is more elusive. Many of the modernist attempts to evoke architectural recognition from those who interact with the buildings fail to achieve their goals. Many buildings remain objects of disinterest, or disdain, or controversy, even decades after their construction. The John Hancock Tower in Boston's Copley Square, by Henry N. Cobb for the firm of I. M. Pei and Partners, is a well-known example. It stands next to H. H. Richardson's Trinity Church (built 1872–7), which draws upon European architecture in obvious ways, adapting traditional, highly representational design elements in variations that seek to establish a distinctly American appropriation of the European legacy. Cobb's 1976 office tower takes an utterly different approach, using a glass curtain wall to mirror the past as it reflects the facades of Trinity. As the skyscraper rises, it reflects the sky and continues the lines of the church's spires into the air. To their detractors modern structures in such situations are absurdly out of place, destroying the integrity of venerable sites, but for their defenders they offer another kind of integrity: the honest assertion that we who visit historical sites inevitably do so as moderns, and see the historic buildings through very modern eyes.

Both modern and postmodern architects have done plenty of explicit referencing of historical architecture. Traditional house styles remain popular

Trinity Church and John Hancock Tower, Boston

for residences, and the New Urbanism embraces these styles fully rather than resisting them. One can think of many architects—such as Philip Johnson, Michael Graves, Robert Venturi, and Charles Moore—who began quoting historical forms profusely in the 1980s, developing complex ways asserting modernism amid historical references, combining and exaggerating decorative elements in a playful and sometimes satirical manner. With such styles it can be difficult to know how genuine the historical interest of the architect might have been. The buildings can play to the traditionalist sensibilities of the public even as they signal their modernism to the modernists. But one wonders whether what is sought is a synthesis of the modern and the historical or rather an escape from the expectations of both.

Something very different, and much more obviously hermeneutical, is at work in the connections to history that Steven Holl established in his 1997 Chapel of

Michael Graves, Portland Building

St. Ignatius at Seattle University (Holl 1999). One of Holl's goals in the project was to humanize the modern without departing from it. The building's program invited a design that was rich in symbolism and reflective of a long Catholic heritage, both in ritual and in architecture. The resulting building's modernism strikes one immediately, with the minimalist rectangles of its footprint forming a simple enclosure, and its geometry repeated in a reflecting pool. The restraint of the concrete walls balances the sculptural freedom of the rooflines, creating a dynamic tension even in one's first impressions of the building. Holl produced watercolor sketches that imagined the experience of moving through the building before he developed the floor plan, thus giving priority to the bodily relationship to the building and its perspectival unfolding as one explores it. The human hand is evident in the incorporation of many crafts: hammered cedar for the entry doors, alter, and baptismal font; hand-textured plaster for

Steven Holl, Chapel of St. Ignatius, Seattle University

Chapel of St. Ignatius   Chapel of St. Ignatius

the walls, and blown glass for the lighting. One's gaze upon surfaces in this space begins to feel like an encounter with persons. Historical associations may not strike one immediately, but emerge gradually as one interacts with the building. In the stain of the concrete one can see the warm tints of the stone in Roman churches, which are also recalled through touch in the heavy exterior doors. The judicious use of colored glass, combined with hidden colored surfaces that reflect it, creates auras of harmonized color that give distinct personalities within different areas of the open plan. As colors drift in the air they recall the hushed atmospheres of pre-modern churches, even as the shapes of the colors in a window, or splashing on a wall or floor, may resemble an utterly modern Suprematist painting. While the arching forms of the ceiling never overtly imitate traditional vaulting, they reach upward with the same aspirations as those arches. While the legs of the altar seem idiosyncratic from most angles, from straight on they take the shape of an alpha and an omega. In such examples the presence of history is felt as a kind of resonance within the modern. Recognition of historical connections emerges as a dimension adding to the depth of the work. This depth dimension unfolds progressively as the details of the building become familiar. In such moments of unfolding recognition one has the experience of discovering countless marks of the past upon the present. It forms an analogy, or an allegory, of an emerging consciousness of the effects of history.

In such moments of unfolding recognition one has the experience of discovering countless marks of the past upon the present.

The abstraction of forms in this building succeeds, as the best of abstract art does, in heightening the multivalance of imagery in the work. In part this richness results from an emphasis on sensual qualities. Every surface has a personality that contributes to the vibrant quality of community in the space. Crucial to this effect is the fact that the design of the work, from start to finish,

Holl, Chapel of St. Ignatius

Holl, Chapel of St. Ignatius

was organized around narratives and metaphors. The image of differently colored glass vessels in a stone box was an early inspiration, as was the classic symbolic quality of light from above. One finds oneself, for example, when inside the building, making aquatic associations, with light shimmering on the polished concrete floors like the surface of a pool and the sweep of rising white walls like billowing sails. Certain design features draw, in largely suggestive ways, upon imagery from the life of St. Ignatius and the imaginative exercises that were part of his spiritual teaching. By such means the chapel connects not only to visual and architectural traditions, but to scriptural and poetic ones as well. As these allusions connect with a Western and Catholic tradition, the meditative spaces of other sacred traditions are not far off. The placing of single stone within the spare geometry of the reflecting pool recalls the Zen rock garden, for example. By such means many venerable histories reverberate through this very "non-traditional" place.

CHAPTER 4

Humanism in the Age of Science

As the seventy-year-old Socrates sat in a prison cell in the ancient city of Athens, awaiting execution at the hands of the state, he reflected with some of his closest companions on his own intellectual biography. In Plato's depiction of the story, Socrates describes how, as a young man, he had been passionately interested in the wonders of nature, and had read avidly in the books of the sages who explained the innermost workings of the universe. But he had soon become frustrated with these thinkers, for the universe they described was a purely mechanical system. In this mechanistic cosmos human understanding and intentions had no explanatory power; the good and the beautiful had no coherent place. Even Anaxagoras, who said that "mind" steers all things, turned out to elaborate this assertion with an array of mindless mechanical operations. Socrates was in prison because an Athenian jury had condemned him for impiety and for corrupting the youth of the city, but he was also there because he had refused an escape plan that his friends had arranged for him. Yet how would one of the naturalistic thinkers explain his actions?

> [He would] say that the reason that I am sitting here is because my body consists of bones and sinews, because the bones are hard and are separated by joints, that the sinews are such as to contract and relax, that they surround the bones along with flesh and skin which hold them together, then as the bones are hanging in their sockets, the relaxation and contraction of the sinews enable me to bend my limbs, and that is the cause of my sitting here with my limbs bent … but he would neglect to mention the true causes, that, after the Athenians decided it was better to condemn me, for this reason it seemed best to me to sit here and more right to remain and to endure whatever penalty they ordered. For by the dog, I think these sinews and bones could long ago have been in Megara or among the Boeotians, taken there by my belief as to the best course, if I had not thought it more

> right and honorable to endure whatever penalty the city ordered rather than escape and run away. To call those things causes is too absurd.
>
> (Plato 1997, 85 [98c–99a])

The story was a favorite of Gadamer's and allusions to it appear often in his writings. He notes that Socrates describes his alternative strategy for understanding the world as one of turning away from the mechanical approach and toward the ways that things are spoken of in language—what Plato scholars call the "turn to the *logoi*." The heritage of words contains the whole gamut of explanatory meanings—from the purely physical to the uniquely human, and from speech about human experience to the mysterious speaking of the divine (Gadamer 1980, 32–5; 1986b, 24–6; 1990b, 456; 1998, ch. 5).

The attraction that inquirers have to both nature and language shares a single motivation: to grasp the patterning of things that emerges in experience. Why do stars move in the patterns that they do? Why does fire burn the same way all over the world? Why do we say that justice should apply equally to everyone? Understanding is always a matter of grasping patterns, and without patterning nothing could be understood. In this fact one sees again a positive role for abstraction. The words and symbols that identify patterns necessarily abstract from their concrete embodiments. Something of the concrete being of the phenomena is lost in the process, but what is gained is a degree of understanding that is not obtainable by any other means, and the worded character of this understanding allows it to be communicated.

The attraction that inquirers have to both nature and language shares a single motivation: to grasp the patterning of things that emerges in experience.

But the autobiography of Socrates also expresses grave doubts about certain kinds of abstraction and their consequences, doubts which persist for Gadamer in the context of latter-day scientific culture. One concern is with the abstraction present in the mechanistic assumptions of naturalistic accounts of phenomena. Most importantly, for Socrates, mechanical explanations make nonsense of moral reasoning and judgment, yet these are the most desperately needed forms of thinking for human life. For Gadamer such doubts regarding the limits of scientific reason have not lost their relevance in the millennia that have passed since the days of Socrates. While the science that we pursue today has become immeasurably more sophisticated, the philosophical concern remains: can the methods of natural science claim to speak with authority about human life? Should the kind of instrumental reasoning that science uses to produce technologies also be used to organize human interaction? Gadamer's inclination is to answer both of these questions in the negative. The narrowing of concern and the objectification that takes place in natural science limits its explanatory power when applied to human affairs, and when society is rationalized on such an objectivistic basis the results pose serious threats to human autonomy and freedom.

Such concerns have parallels in architectural theory. The emergence of architectural modernism was driven by technological innovations in materials and engineering techniques, to the point where mainstream views began to see the task of design as one of accommodating the aesthetics to the technology rather than the other way around. But this movement raised a host of persistent questions. For example, does the emphasis on the functional, mechanical nature of architecture, with its pervasive symbolization of the technological revolution *per se*, break architecture's ties to the humanities, and perhaps to the human dimension in architecture's historical identity? Allied with this question is an architectural version of the question of social engineering. Architecture has been recruited on a grand scale in the reshaping of human communities, and architects have produced grand visions of the rationally organized city. But has architecture thereby served its proper calling, or has it been swept up in the romance of an obsessively technological age?

Humanity in the emergence of the natural-scientific ideal

The philosophy of form that Plato and Aristotle developed to challenge the mechanistic philosophies of their day was challenged, in turn, by the resurgence of natural-scientific thinking during the Renaissance and Enlightenment periods of European history. Very legitimate charges were leveled by writers such as Galileo, Descartes, Bacon, and Thomas Hobbes against the dogmatism, faulty observation, lack of experimentation, underdeveloped mathematics, and naïve anthropomorphism of the scholastic thinkers who continued to ground their science in the teachings of the ancient Greeks. But these critiques also indulged in polemic exaggerations, as in their expressed desire to tear down the whole of Western learning and begin anew (Bacon 1960, 106–10; Descartes 1993, 4–5). The rejection of anthropomorphism in scientific explanations is a wholesale affair, setting up the conditions for "human" and "scientific" to become opposites by definition. But these polemics can mask the fact that Bacon and Descartes also drew heavily on earlier traditions for their visions of human excellence and worth. They want to replace philosophy with science, yet at the same time they speak like the older philosophers when they paint their vision of the god-like powers of the human intellect and the moral imperative of technological advancement.

> The rejection of anthropomorphism in scientific explanations is a wholesale affair, setting up the conditions for "human" and "scientific" to become opposites by definition.

Positivist schools of thought in the nineteenth and twentieth centuries resolved this ambiguity in favor of natural science. "Positivism" is so called because it takes positive science as the sole or ultimate rational authority. Formal mathematics and empirical method are considered the only means of obtaining reliable answers, with the implication that any questions that science cannot address simply cannot be reasonably addressed. Among these are questions of moral value, the ultimate nature of reality, and all matters pertaining to spirituality and divinity. Positivism is embraced, of course, with varying degrees of consistency, so that one frequently finds today scientists who do not hesitate to explain every sort of human activity in terms of highly speculative neurobiology or evolutionary psychology. Contemporary positivism often retains the claim that only natural scientific reasoning is reasonable, but it has become much bolder in eschewing any limits on what natural-scientific reasoning can do.

Gadamer's thought reflects influences by a number of thinkers—including Kant, Hegel, Husserl, and Heidegger—who respected modern science while rejecting claims for its exclusive authority. Kant argued that a deterministic Newtonian universe is compatible with belief in the rational integrity of human moral principles. Hegel made one of the last great philosophical efforts to integrate metaphysical logic with the structures of nature revealed by science. Husserl critiqued the unquestioned epistemological assumptions of positivism in order to recover the scientific integrity of philosophy. Heidegger, who always recognized a place for science, challenged the totalizing way in which technology had "enframed" the world, and sought to counter this effect by recovering the poetic foundations of human dwelling on Earth.

Gadamer bears the stamp of all of these philosophies, yet his thinking is never simply identical with any of them. For most of his career Gadamer's chief complaint against natural science was that its methodology excludes of any kind of hermeneutic awareness, an exclusion which it combines with attacks upon the rationality of the disciplines that *do* employ a hermeneutic approach. Against positivism Gadamer positions himself as a defender of the rationality of the humanities, showing, in myriad ways, how their wisdom has been underestimated. From this point of view Kant is seen to grant too much authority to Newton; Hegel is too optimistic in seeking the marriage of natural science and speculative philosophy; Husserl wanted too badly for phenomenology to resemble a rigorous science. Even Heidegger, to whom Gadamer is arguably closest on these matters, is limited by his propensity to locate the alternative to scientism in a rarified thinking of being that polemicizes against elements of the humanistic tradition that it would do better to retrieve (Gadamer 1976, chs. 8 and 9; 1992, 128).

In an afterward to *Truth and Method*, and in later reflections on the path of his thinking, Gadamer acknowledged a certain one-sidedness to his criticism of natural science. So eager had he been to point out the lack of self-reflectivity in the application of scientific method that he had failed to acknowledge fully the legitimacy of science within its proper sphere (1990b, 551–61; 1997, 40–41). Yet Gadamer never ceased to characterize his fundamental role as that of an advocate for the humanities and their hermeneutically recoverable potential in the face of the dominance of natural- and social-scientific methodologies. In an essay in which he acknowledges the limitations of his earlier thought on science, he also writes,

> precisely because of the nonunifiability of philosophy with the modern natural sciences, hermeneutics as philosophy does not have as much to learn from the theory of modern science as it does from the older traditions, which it needs to call back to memory.
>
> (1997, 30)

Modernity and the ethical function of architecture

The impact of the rise of natural science upon the practice of architecture has been a central interest of architectural theorists who have drawn upon Heidegger and Gadamer. The "divided representation" that Vesely analyzes, for example, has its roots in this history. As architectural interpretation and education came to be captivated by the purely geometrical qualities of structures; as breakthroughs in engineering brought with them their own ideals of simplicity and efficiency; and as space, time, volume, perspective, and optics came to be viewed in uniformly quantitative terms, the symbolic and poetic dimensions of design acquired a superficial status—both in the sense of being pushed to the exterior surfaces of structures and in the sense of being non-essential. Vesely and Pérez-Gómez have identified a decisive moment in the emergence of the division in early nineteenth-century growth of positivism as a fundamental world view. Pérez-Gómez also provides detailed treatments of the pre-history of the movement toward positivism going all the way back to early debates over the Vitruvian orders. In his view, what has been lost, as positivistic scientism has come to dominate the culture of architectural practice, is what was most central to architectural interpretation in the age of Vitruvius:

a humanistic and poetic significance—with roots deep in traditions of myth and cultural narrative—that guides the design process and suits the building to its purpose and place (Pérez-Gómez 1983). For Pérez-Gómez, as for Gadamer, the older mythologies and rituals must be called back to memory—not in their own context, to be sure, as if one could turn back the clock and forget modernity—but as poetic elements in our tradition that can ring with new brilliance in the contemporary cultural landscape. The aspiration to build today can be inspired by *eros*—a desire born of the sense of one's incompleteness—no less than in any other age in history, and in the community of desire that unites creative spirits throughout the world and across historical ages there may be found the antidote to the fragmentation that characterizes architecture today (Pérez-Gómez 2008).

The aspiration to build today can be inspired by *eros*—a desire born of the sense of one's incompleteness—no less than in any other age in history.

In Karsten Harries' terms, what the combination of aesthetic superficiality and positivism has wrought is the loss of a sense of "*ethos*" in architecture—the rich blend of qualities that give it normative character and interpretive power. While architectural *ethos* in this sense is broader than "ethics" in the sense of moral philosophy, Harries shows how, at times, normative forces in architecture do take on the character of a moral imperative. As Bacon and Descartes promoted modernism as a way out of blinkered traditionalism, so Adolf Loos and Le Corbusier rejected the ornamentation of traditional styles as outworn symbols of elitism and decadent fantasies (Harries 1997, chs. 3 and 4). Modern styles, with their unprecedented capacities for efficiency, repeatability, and scale, introduced the possibility that every member of society could live and work in sanitary, light-filled structures. As the public learned to appreciate, in the design of these modern buildings, the streamlined forms that they already admired in the styling of automobiles, airplanes, and ocean liners, their aesthetic sensibilities would come to harmonize with the progressive social ideals of the modern age.

By emphasizing this moralistic dimension of modernism, Harries' analysis reveals a tension in architecture's current situation. The ethic of modernism creates profound barriers to any attempt to simply return to historical styles. Every such attempt must face charges, not merely of aesthetic nostalgia, but of a socially irresponsible backwardness. Yet at the same time the modernist ethic, as it has unfolded in practice, has failed to realize the harmony that it sought. Much of the public is unable to find in modern architecture the range of meaning that its early advocates saw in it. The technological imperative, which has in many respects become an end in itself, can seem, like the fanciful constructions of the ancient atomists, to subordinate uniquely human purposes to the demands of the great cosmic machine.

It is fair to say that in the attempts by contemporary theorists to make sense of the complex combination of motives in modern architecture, Le Corbusier retains a unique status. While he wrote some of the most strident manifestoes against traditionalism in architecture, his radically modern works were at every moment inspired by a normative vision of human significance. As much as his designs celebrate the triumph of modern engineering, they equally reflect the visual sensibilities of a painter, the spatial imagination of a sculptor, and the symbolic ingenuity of a poet. Norberg-Schulz, who writes in very negative terms about the loss of place in modern architecture, nevertheless returns regularly to the example of Ronchamp to show how the meaning-function of architecture can be carried brilliantly into the modern idiom (Norberg-Schulz 1979, ch. 8; 1985, 88; 2000, 107–10). However, he considers Le Corbusier's particular solutions difficult to repeat. The visions of Alvar Aalto and Louis Kahn, which have ties to Le Corbusier, turned out to be alternatives that were more easily communicable to students and fellow architects.

Harries takes the multiple facets of Le Corbusier's vision as evidence of a fundamental tension within its approach to functionality. On the one hand, Le Corbusier sees the engineer's aesthetic as being far ahead of the architect's, and yet architecture cannot simply be handed over to engineering; architecture must serve something higher than utility (Harries 1997, 235–6; Le Corbusier 1986). Pérez-Gómez offers a detailed reflection on Le Corbusier's monastery,

Sainte-Marie de la Tourette, praising its ability to achieve poetic resonance in an utterly modern idiom.

> Architectural meaning appears in *our* intersubjective space as in the space of a metaphor, as a recognition of that which cannot be reduced to words and yet begs to be named. La Tourette still proclaims the possibility of participatory, non-reductive representation in the world of electronic media and simulation; the possibility of a culturally significant architecture ...
> (Pérez-Gómez and Pelletier 1997, 368)

Examples such as these could be multiplied because any author concerned with meaning in modern architecture can hardly afford to ignore Le Corbusier. His work stands as a permanent opening to a world of possibilities. They are monuments to an utterly radical, yet uniquely integral vision for architecture. It is hard to regard the vision a straightforward solution when it exhibits so many internal tensions and when its attempt to mainstream the brutalist aesthetic has met with so much public ambivalence. But many of Le Corbusier's buildings illustrate well the qualities that Gadamer attributes to a "classic" work. One can return to them again and again to find further, unsuspected relevance and inspiration. Their limitations and failures are often as illuminating as their successes, and all of the critiques that are mounted against them tend to do more to increase interest than to diminish their status.

Many of Le Corbusier's buildings illustrate well the qualities that Gadamer attributes to a "classic" work.

Hermeneutics and the rationalization of society

One must credit Jürgen Habermas with bringing Gadamer's hermeneutics into the discourse of social and political theorists. In his early but still influential book *The Logic of the Social Sciences* (1988) a central concern of Habermas is the problem of relating the third-person accounts human life that are the product of social sciences with the kind of first-person accounts that one finds in the humanities, the arts, and traditional historical narratives. What I mean by "third person" here is an approach that seeks patterns of social interaction which exhibit their regularity quite independently of the intentions of the persons involved. As with the natural sciences, the patterns that social-scientific method aims to reveal may be more or less invisible to ordinary experience and may thus be masked in first-person reporting. This goal of achieving a third-person perspective has become, in the social sciences, almost identical with the idea of objectivity, and the vast majority of articles in social science journals take great pains to appear as if their research was conducted in the manner of a controlled scientific experiment.

But Habermas identifies a number of problems with the attempt to study humanity on the model of empirical natural science. In the process of imitating the chemist or biologist one tends to downplay the inevitable role of interpretation in framing questions, constructing instruments for data gathering, and making sense of the results. One has difficulty connecting one's third-person results with first-person experience because one cannot really get past the pejoratively subjective status that was accorded that experience from the outset. Thus one filters the data through an objectivistic interpretive framework that normally does not (and apparently cannot) call itself into question, because that would be less than "scientific."

Gadamer's hermeneutics, argues Habermas, can temper the positivistic excesses of social science by pointing to the inevitably ubiquitous role of interpretation in the entire endeavor. Gadamer charges that no method, no canon of empiricism, can free one from a cultural horizon or from history. Habermas likes the fact that even as Gadamer asserts the impossibility of shedding a first-person perspective he simultaneously claims that there is much to this personhood that functions quite apart from one's subjective intentions. In this respect Gadamer is in harmony with the social scientist's interest in what people do apart from their knowing and wishing. Habermas also appreciates how this happening of history within us is said to take place by means of language. He is drawn to language philosophy over phenomenology because of the inherently interpersonal and quasi-objectified nature of language (qualities that

Habermas explains by drawing on Ludwig Wittgenstein as much as Gadamer). Gadamer's hermeneutics, then, has value to Habermas not as a polemic against social scientific method but as an accompaniment to it. Hermeneutics provides a bridge from the historical as analyzed to the historical as reflectively experienced. It raises questions about the cultural embeddedness of theory that every theorist should bear in mind. In fact, Habermas goes so far as to suggest that it is in this complementary role that philosophical hermeneutics can *best* flourish (Habermas 1988, ch.8).

<u>Gadamer's hermeneutics, argues Habermas, can temper the positivistic excesses of social science by pointing to the inevitably ubiquitous role of interpretation in the entire endeavor.</u>

But there is a certain backhandedness in this appreciation of hermeneutics, because Habermas reads Gadamer's "rehabilitation" of prejudice as simply making the case for the authority of tradition. Hermeneutics is best nested within social science, according to Habermas, because it lacks any independent means of critiquing tradition. Habermas famously invoked, as an alternative to the Gadamerian's reverence for the wisdom of tradition, the model of the Freudian analyst's interpretive approach to the first-person accounts of his or her patients. The analyst becomes immersed in the patient's linguistic world, but always with a critical perspective. The analyst is looking for systematic distortions in the patient's descriptions of his or her experience; almost nothing is taken purely at face value (1988, ch. 9; 1990a; McCarthy 1978, 178–213; Warnke 1987, ch. 4).

There is much in this Habermasian reading of Gadamer that is simply mistaken. The most common Gadamerian response to the reading is defensive: there is quite emphatically a critical dimension in philosophical hermeneutics. Tradition is a legacy of questioning, doubting, and criticizing as much as it is a legacy of beliefs. Hermeneutics makes the case for a critique of cultural dogmas that is founded on understanding rather than on some scientific algorithm or some misplaced faith in the vantage point of the present. It is true that for Gadamer prejudice conditions one's initial opening to cultural meaning, but Gadamer is not thereby saying that one should end where one begins. Quite the contrary. It is the very core of the hermeneutic project to pursue questions with the knowledge that genuine understanding will alter the assumptions with which one began.

These assertions put in a nutshell the now standard defense of Gadamer, but one can put him on the offensive as well. Habermas does not seem to fully grasp that for Gadamer the reasoning of the social sciences is an extremely narrowed form of hermeneutical and dialectical thinking. If Habermas understood this he would never suggest that the broader kind of reasoning should be nested in the narrower. Habermas acknowledges that the social sciences seem unable to question their own interpretive frameworks, but he somehow does not see this as a threat to their critical authority. But in Gadamer's view the functioning of scientific method here is altogether too magical: social scientists believe that the canons of empirical method overcome the problem of bias, but in fact the use of empirical method instead of hermeneutics deprives social scientists of the ability to question their own starting point, whereas that kind of questioning is, for Gadamer, the very key to meaningful critique.

So from Gadamer's point of view, not only does Habermas fail to go far enough with hermeneutics to discover its critical dimension, the hybrid that Habermas proposes, in his laudable attempt to escape the excesses of instrumental reason, grants too much to that instrumentalism. He wants to mitigate the influence of instrumental reason in social inquiry but he ends up largely endorsing it; he wants to oppose social engineering but he cannot help justifying it (Gadamer 1990a; 292–3). In this light his embrace of Freud, in particular, seems problematic, for Habermas is simply assuming that there exists a person or persons who possess the kind of advantaged perspective on culture that is the equivalent of the Freudian analyst, but who would that be? And by what means could it be known that this perspective is the definitive critical one? These questions are hermeneutical questions (1984c, 78–9; 1990a; 279–81; How 1995, 174–7). Indeed, in the Freudian analogy, I would argue, one can also see something of the paternalism of the social engineer, particularly if one looks at the situation from the perspective of the patient rather than that of the analyst. In Freudian psychoanalysis the patient is not allowed to interpret but must only provide materials for the interpreter. This interpretation is done according to a codified interpretive framework that the patient has no power whatsoever to affect. If, from the analyst's point of view, the procedure is critical, must we not say that from the patient's point of view it is the very epitome of a controlling use of reason? For a Gadamerian, instrumental reason is only genuinely put at bay when the intelligence and agency of those who are to benefit from social organization and change have a dominant position in the process. The Freudian analogy does not serve that goal well.

For a Gadamerian, instrumental reason is only genuinely put at bay when the intelligence and agency of those who are to benefit from social organization and change have a dominant position in the process.

Social engineering and the city as work of art

Besides his iconic individual buildings, Le Corbusier is best known as the originator, in the 1920s, of the idea of the "Radiant City" (1967). The model formed a thoroughly modern conception of the urban good at many levels. It sought to bring the beauty and efficiency of life on an ocean liner to residential buildings. It aimed to compete with the suburbs by building into the air and freeing up most of urban land for parks and recreation. It aspired to social egalitarianism through the modular fabrication of sanitary living spaces that anyone could afford. It planned to do away with traffic congestion by putting all of the cars on expressways. It was a re-imagining of the city from the ground up and down to the last detail. Rationalized planning was to bring order to the maddening chaos of cities. Modern technology was to be pushed to such an extreme that something of its opposite could emerge: the greening of the city. Here again, Le Corbusier can be credited with producing an ideal that is utterly technological yet artistically inspired at every turn. His "Voisin Plan" for the Left Bank of Paris remains one of the most astonishing projects ever conceived—astonishing in its scale, in the uncompromising thoroughness of its modernism, and in its gleeful destruction of huge tracts of property in one of the world's most storied cities.

Across the globe, following World War II, at least some of the elements of the Radiant-city vision were embraced for mixed-income housing, and more predominantly for low-income projects. In New York City, from the late 1940s to the early 1960s, the great American master-builder Robert Moses—one of the most powerful public officials the country has ever produced—oversaw the creation of enough dwelling units to house more than a quarter of a million people, much of it done on the Radiant-City model of massive superblocks with uniform high-rise buildings. But this vision has come to be seen as a flawed effort and a paradigmatic example of the inherent limitations of large-scale single-use planning projects.

One of the most effective critics of the Moses vision for New York was Jane Jacobs, whose polemical book *The Death and Life of Great American Cities* has been called a classic of American letters (Jacobs 1993; Fulford 1992). It attacks the Urban Renewal vision at several levels, making its case for a different kind of urban vision with eloquence and extraordinarily insightful common sense. In her attacks, Jacobs represents the power that concrete, situated thinking can exercise in opposition to abstract instrumentalist thinking. In many ways, the instrumental thinking of the planners did not seem interested in understanding the lives of the people it most affected. The concept of "slum" was fashioned with abstract categories such as residents per acre and building age. There was little investigation into the quality of life and the health of communities in these often lively, frequently ethnic neighborhoods. City officials exaggerated the amount of relocation assistance that it was actually offering as they bulldozed these neighborhoods, and the new housing that was put in its place employed an utterly different conception as to how a city is organized and how it works.

The new urban vision, born of Garden-City and Radiant-City thinking, could not reproduce the local urban vitality that it replaced because its authors could not see how dependent a city is upon mixtures of urban uses. The superblock model is based on the separation of uses. Dwellings are concentrated in single-use high-rise buildings, separated from work places, retail shops, and cultural establishments. The old neighborhoods, when they succeeded, did so by combining all of these uses in a walkable arrangement that depended heavily on public streets and sidewalks. The failure of modernist planning to produce

North End, Boston

vitality goes back, in part, to its habitual abstraction, its tendency to think of the city as a set of two-variable problems—relating residences, say, to workplaces. But every use that urbanites make of their city is a solution to a multivariable problem; they are always doing many things at once. One cannot deduce their behavior, then, from a set of categories; one must come to recognize patterns as they emerge and figure out how to avoid inhibiting them (1993, ch. 22). In this connection Jacobs' view poses particular challenges to urban designers, charging that "the city cannot be a work of art" (1993, 485). It is not something to be laid out on a drafting table, finding an aesthetically gratifying way to organize the lives of citizens. A city thrives on the freedom of people to make their own way of life and to realize their own creative aims. To be granted an aesthetic solution is to be deprived of that freedom, whereas to foster creative freedom requires a certain amount of disorganization and lack of definition.

A city thrives on the freedom of people to make their own way of life and to realize their own creative aims.

I am describing the much-discussed historical example of the Jacobs–Moses conflict not to provide an interpretation of it (for that is far too large a task to attempt here), but simply to indicate the relevance of philosophical hermeneutics to the case. The most obvious relevance is in its opposition to overtly intentional social engineering. But more pointedly, Jacobs unearths elements of social control that were operative beyond the intentions and the self-understanding of its practitioners. In Gadamer's terms we would say that the planning mentality functioned as a horizon that prevented a whole host of pertinent questions from being asked. Even the language that was used by Urban Renewal advocates, such as the preference for the term "residential zone" over "neighborhood," limited the types of answers that could satisfy the formulation of the problems.

More challenging still is the possibility that the entire field of urban design is afflicted with horizontal blind spots that make its products more controlling than they ought to be. This challenge has been the source of many ironies in the repercussions of the revolt against Urban Renewal. Many New-Urbanist projects, for example, have produced neighborhoods or new towns out of whole cloth, designing into them the very sorts of mixed use and walkability that Jacobs advocated. Yet in the comprehensiveness of their design and the unity of their architectural styles these projects would seem to violate Jacobs' strictures against turning the town or city into the designer's work of art. To really belong to the citizens, these places must become, in some fashion, the *citizens*' work of art. Urban design, as Gadamer said of all architecture, performs its artistic function, in part, by artistically receding, making a meaningful place for creativity to happen. Buildings actualize human meanings, but more essentially and more importantly they make meaning *possible*.

Implicit in examples such as the Urban Renewal debate is a host of questions pertaining to the relationship between theory and practice. Jacobs' work is frequently criticized for lacking an adequate theoretical foundation; Jacobs, for her part, charged that the politicians and planners who were trying to remake New York were so besotted with theory that they could not see the vitality of the communities that they were about to dismember. But what, after all, *is* theory? And from what does the authority of practice derive? These questions draw our attention to another defining feature of Gadamer's philosophy and another distinct type of relevance that it has for built environments. It is the feature to which Gadamer refers when he locates philosophical hermeneutics within the tradition of "practical philosophy."

CHAPTER 5

Practical Wisdom in Creative Collaboration

Gadamer wished that his hermeneutics would be seen as a form of practical philosophy, joining a tradition of practical thinking that goes back to Socrates and Plato. From the fact that Gadamer calls his philosophy "practical" one should not infer that he simply prefers common sense to theory. Rather, philosophers in this tradition typically regard theory as a needed refinement of practical know-how. Gadamer himself does so when he refers to theory as a highest form of practice (1986b, 175). Such a statement endorses the value of theory, but it contrasts sharply with a common conception of theory as a set of principles worked out in the abstract and then "applied" to practice. For Gadamer, to isolate theory in an abstract realm, away from the practical insights of lived experience, is to rob inquiry of its ability to think hermeneutically.

The practical wisdom tradition

In his efforts to relate theory and practice Gadamer once again finds a touchstone in the nuances of the Platonic *Dialogues*. The people that Socrates engages in conversation in these works are often men of practical accomplishments who are considered by many to be "wise" for possessing highly useful skills. But the questions that Socrates asks reveal limitations in the scope of their wisdom. For example, two famous generals in the dialogue *Laches* can define the term "courage" as it is understood in the circumstances of battle that are familiar to them. In this context it makes perfect sense to say that courage is to be found in "the man who fights the enemy while remaining at his post" (Plato 1997, 676 [191a]). But the generals have difficulty making sense of the term in special military circumstances, such as a strategic retreat (where the point is precisely not to remain at one's post), or in non-military circumstances such as courage in the pursuit of truth. In alluding to Platonic examples of this sort Gadamer wants his readers to notice a tension in the discourse. Socrates' search for a comprehensive definition pushes practical wisdom to a higher level of generality than it would otherwise attempt (Kidder 1995). Socrates' movement toward theory is an organic development of practical wisdom, not an alternative to it.

Socrates' movement toward theory is an organic development of practical wisdom, not an alternative to it.

An equally important ancient source for Gadamer's thinking on this subject is Aristotle's *Nicomachean Ethics*. Aristotle did not interpret Plato as a practical philosopher in the Socratic style, but as someone inclined to divorce theory from practice. There is no doubt some fairness in this characterization, in that Plato founded his famous Academy so that young men could cultivate the intellectual life outside of the turmoil of politics, establishing a curriculum that included such abstruse subjects as mathematics. Aristotle took Plato's own teaching in the Academy to be positing an abstract, mathematized reality as superior to the world of ordinary experience, and he saw this metaphysical approach as shifting Plato's ethics away from practical life and toward a rarified pursuit of "the good" that lies beyond the world of experience. Whether this characterization is fair or not, Aristotle positions himself as thinking counter to this tendency, presenting his own ethics as development in practical wisdom (Gadamer 1984c, 115–17; 1990b, 312–18; 1998a, 16–20).

The goal of this development is to find measure among competing desires and to judge fairly in complicated circumstances. Aristotle's ethics employs principles, to be sure, but they are meant to be used as guides to practical decision-making by persons of well-cultivated character. Justice, for example, includes deciding fairly in the distribution of goods. That is to say, justice implies a principle of equity. But what a just distribution means concretely will depend on the persons involved and their circumstances. One would not divide food equally, Aristotle notes, between an accomplished athlete and a beginner, because the same amount might seem to be too little to one and too much to the other (Aristotle 1999, 24 [1106b]). Here deciding justly requires knowledge of principles, but equally requires familiarity with the circumstances in which the principles are to be realized. One can recognize in this example the kind of reasoning that Socrates was seeking. It is a type of practical reasoning ("*phronesis*" in the Greek) that goes beyond practical skill (or "*techne*") that arises out of experience alone. It is a reflection on experience that moves towards generality without losing sight of particulars. It is oriented to what is common in different cases, but without simply ignoring differences among particular cases (Gadamer 1990b, 314; Risser 1997, 110–14; 2002).

The tradition of practical philosophy survives in later theories of hermeneutics, according to Gadamer, by virtue of the problem of "application." In Christian worship, for example, the challenge of the sermon or homily is to show how the practical meaning of the biblical text is to be realized in the very different life of the contemporary community of believers. Application in such cases is not incidental to the meaning of scripture because the whole point of scripture is to inform a way of life. In the juridical context, "application" similarly determines how laws are to be made sense of in terms of contemporary society and how they are to be realized in particular cases. Laws cannot themselves determine how they are to be applied. The practical wisdom of judges, attorneys, juries, and legal scholars is critical in such decisions. Both scripture and law, then, can only be made actual and relevant by means of cultivation in *phronesis*—practical reasoning—on the part of those who have the responsibility of applying them (Gadamer 1984c, 126–8; 1990b, 309–11).

With the emergence of natural-scientific paradigms in the study of society and culture the distinction between theory and application that one finds in the traditions of scriptural and legal hermeneutics is pushed to an extreme. On the model of mathematics, theory becomes the pure relating of abstract concepts; on the model of empirical study of nature, theory becomes detached observation that looks, in a neutral way, for laws of human nature and interaction. Our look at the Gadamer–Habermas debate showed how Gadamer's philosophy seeks to recover the interpretive dimension from its eclipse in positivist social science, but now it should be noted that in so doing philosophical hermeneutics elevates the rationality of practitioners who reflect on their practice and who seek to improve that practice through reasoned discussion with many kinds of dialogue partners.

Hermeneutics in practical deliberation

It is not only in the circumstances of law or religious life that one must make practical decisions in light of rules, teachings, codes, customs, and historical precedents. Any important decision will have some role for these considerations; every decision will be, in some sense, an application. Cases involving the interpretation of historical texts or achieving understanding across nations and languages are simply the more dramatic instances of hermeneutic encounter. Any decision that is made in collaboration with other people can benefit from the kind of understanding that involves hermeneutic awareness. Often deliberations are focused on simple practical decisions, and in these cases a limited use of common sense or instrumental thinking may be sufficient, but when the problem is complicated, involving multiple stakeholders and many kinds of interests, then the processes of reflection, discussion, deliberation, and decision will manifest more strikingly their hermeneutic dimension. For, strictly speaking, no one is ever in exactly the same horizon. Every attempt to work together, to make decisions together, to live together requires complex, often difficult efforts of understanding. To grow in one's realization of the need for understanding, to learn patience with the effort and to develop a sense of its own intrinsic rewards, to have tested principles in wide ranges of circumstances, to have called them into question and to have revised them where needed, to have built trust and to have formed friendships even within the context of contentious deliberations—all of these are signs of development in practical wisdom, a development which is more determinative of one's character and identity than any skill or set of convictions could be.

> Any decision that is made in collaboration with other people can benefit from the kind of understanding that involves hermeneutic awareness.

As a practical philosophy, hermeneutics does not provide answers to problems so much as it indicates patterns of inquiry that one can undertake in the process of thinking them through and collaborating with others to solve them. These patterns come into play whenever we say that members of a deliberating group are trying to "listen to one another" or to "understand one another's point of view." In such moments the purpose is not simply to state the principles or policies at stake but to try to grasp the different ways they are being interpreted by the group's members. These interpretations bring ideas to bear, but even more so, histories: the history that has lead up to the decision-making circumstances, the history that shapes the roles of the group's members, the history of their interactions with individuals and organizations that determines the levels of trust and confidence that they bring to the table, the histories of fairness or unfairness that define the members' roles as defenders or activists, and the symbols and feelings that go with all of these histories. To "listen to" others in such circumstances is to open the door to the relevance of all of these histories. A deliberative process that opens this door becomes a longer and more complex one, but one that can yield a more richly informed decision, often building relationships of good will along the way.

Greater understanding in such cases is a combination of recognizing difference and seeking agreement. If one were not interested in agreement then one would not bother; but if there were no difference to recognize then the effort would be unnecessary. Of course, in any group process, there is the danger that "figuring out what we can all agree upon" will narrow the scope of the deliberation, will encourage too much compromise of principle, or will reinforce shared biases of the group (Warnke 2011). But the open-mindedness that is expressed in the desire to understand should not be taken as the willingness to surrender one's position or to stop bringing principles to bear on the discussion. Nor is it an openness that inevitably gives special privilege to the members of the group. The openness that gives voice to others in the group is the same openness that brings perspectives external to the group into view. Openness of this sort anticipates an indefinite potentiality of reasoned views and their motivations. It is a willingness to recognize in the position of others a comprehensible series of steps that has lead them to where they stand today. Openness to this potentiality is not the same as actual agreement. It may lead to an increase in agreement based on better understanding, but it may also simply shift the participants' sense of where the points of disagreement are to be found.

Researchers who study the process of deliberation in the context of urban planning make relevant observations on the role of mediators in situations where group processes have come to an impasse. Mediators in this role have as their goal, in John Forester's words, "the transformation of adversarial expectations to collaborative explorations" (1999, 101). This exploratory shift, I would suggest, often exhibits hermeneutic characteristics. For example, mediators are encouraged to bring into the discussion the personal and cultural narratives that are behind the participants' positions, so as to get at the sources of some of the most intense feelings involved; but at the same time it is recommended that mediators characterize the task as *listening* to one another's stories, so as not to allow a clash of narratives to polarize the debate. It is suggested that mediators invite a wide range of voices into the debate, trying to loosen the hold of a few who have perhaps deadlocked the discussion, and that a consensus model be used so as to give minority voices more opportunities to speak. Mediators are advised to focus on building relationships among group members because one cannot predict where, in such relationships, sources for advancing the dialogue might arise (Forester 1999; Innes 1996; Margerum 2002). In such recommendations one can see an attempt to balance the general and the particular—the general purpose that is common to the group and the general values that the group is meant to realize, but the particularities of the horizons and relationships that each participant brings to the discussion. It is not amiss to regard breakthroughs in mediating processes of this sort as instances of the fusion of horizons, even though, when compared with fusions across large stretches of the globe or expanses of history, the differences in horizons seem small.

Collaboration in architectural creation

Steven Holl frequently calls architecture "the most fragile of arts" (e.g. 2009, 287). In this statement he has very much in mind the fact that architecture is always the product of a complex set of collaborations. There are collaborations that pertain to the design process, to the financing, and to the contracting; there are collaborations that set any building project within the larger task of planning; and all of this will be nested in still larger political, economic, and cultural interactions. In this complex process there will be participants who believe in the art of architecture and who will work hard to make it a reality, but there are likely to be people who are mostly interested in the square footage, or who have very different ideas about how the project should be organized, or who never wanted the thing built in the first place and still do not see why it should be. The fragility of architecture that Holl has in view can be seen as the fragility of art in the face of collaborators who may care little about art, but the more constructive way to read it is to see the architecture as fruit of the delicate art of collaboration. Such an interpretation refuses to separate the nature of architecture from the practice of architecture. It imagines the work of architecture to be work with persons and worldviews as much as with materials and designs.

In the course of pursuing a project the voices of all of the collaborators will have their say, and the architect will need to interpret what is said. This interpretation must take into account the authority of the various contributors, both in terms of their formal roles and in terms of their grasp of what is involved in the process. The responsibility of the architect to the client has a place of central importance because it is the primary contractual relationship. But the task of delivering "what the client wants" is complicated by the degree to which the client can really understand the process of design and what is possible within the parameters of the project. A further complication arises from the fact that the architect has been hired as a professional and an expert, which implies that the value of his or her work for the client will be determined, to a significant degree, by professional standards of the architect's peers with which the client may be altogether unfamiliar (Schön 1983, 291–2; Sirowy 2010, 197).

The task of delivering "what the client wants" is complicated by the degree to which the client can really understand the process of design and what is possible within the parameters of the project.

Both architect and client have responsibilities to the communities in which they are permitted to work. These are expressed formally in the codes and regulative processes that provide a legal framework for the project, as well as the less formal ways in which community members who are affected by the project are invited to provide input. In the case of each of these stakeholders the architect interprets according to a hermeneutic that balances responsibilities to agents with assessments of their perspectives.

Again, although in the grand scheme of things these differing perspectives operate within a single culture, there is a value to thinking of them as different horizons, for their roots in formative experiences and emotions go very deep. For example, a successful developer has an enormous amount of his or her personal integrity at stake whenever decisions are made that will affect the commercial viability of the project. A community activist has formed an entire vision of the moral life around his or her contributions as a public citizen. For each of these persons there are a great many assumptions that never quite rise to the surface. There are questions that seem important to ask and ones that seem irrelevant. There are powerful emotions expressed by some participants that simply do not stir in the hearts of their fellow collaborators. These are characteristics of the interaction of horizons.

Bill Hubbard Jr. would seem to have something like the notion of horizon in mind when he speaks, in *A Theory for Practice*, of the three different "discourses" of the architect, the client, and the community (Hubbard 1995). He describes these, at one point, in terms reminiscent of Schleiermacher's hermeneutics of parts and wholes. The architect, says Hubbard, comes at a project in terms of parts and wholes simultaneously. The program and its design are developed together, unfolding in mutually important and illuminating ways. The client, however, tends to be focused on the whole of the project. He or she has initiated the process as an investment that requires an acceptable return. Thus for the client the parts of the project—the elements that, for the architect, pertain to the design—are all variables. What matters, above all, is the viability of the whole. By contrast to both of these perspectives, members of the community tend to be parts-focused, because their interactions with the finished product will be centered on particular features that may not be central either to the client's program or the architect's design. The matters of most concern will be the externalities that have impacts on neighbors and other community members (1995, 108–9).

Hubbard identifies these perspectives as "discourses" because he is trying to account for the tendency of people in the process of architectural creation to talk past one another. I characterize his account of these discourses as revealing qualities of horizons, for the discourses reflect orientations with many kinds of roots—linguistic roots, to be sure, but also roots in particular experiences that shape an orientation to experience in general. The depth of these roots can make the task of understanding within collaborations daunting, but progress in the effort highly rewarding.

Design as horizon

A hermeneutical account of any encounter of horizons would be incomplete without focusing on the task of appropriating one's own horizon. In working with multiple horizons, the architect is always working *from* a horizon. The most distinctive feature of that horizon is the decisive role that design plays in it. For an architect design is never simply a skill, or even just an art: it is an entire way of being present to the world and sensitized to its defining qualities. The powerful feelings that the young Gadamer had for the floor in his family home illustrates how profoundly the qualities of designed space can educate one's sensibilities regarding order and value in the whole of life. When Holl describes "archetypal experiences" he identifies moments that may persist as inspirations over many years, moments that can serve again and again as points of reference for the possibilities that design opens up.

Bill Hubbard has such experiences in view as he paints the picture of the architect's world of discourse. He describes such a "design epiphany" in a visit to the Eames Case Study house:

> It was not so much the house that made an impression, it was the way life that was lived there. It was the vases of roses—the old, heavy-fragranced kind, always in a mixture of colors, each rose perfect and fully blown, just on the point of dropping its petals. And it was the tea, served on the terrace from an old-fashioned china pot into delicate cups. And always sweets, but not just any sweets: with the tea there would be a great pannettone, or biscotti, or millefeuilles. Or berries, a very few but each fresh-picked, with thick cream, eaten with big sterling dessert spoons. And all of this while looking out over the tawny grass toward a sliver of the Pacific, the whole moment accompanied by the rustle and smell of eucalyptus trees.
>
> (Hubbard 1995, 3)

A person who has made a life in the world of architecture is someone in whose imagination such experiences loom large. He or she has been moved—perhaps early in life—by the power of design to bring order, or peace, or fascination to life. In design the world that ordinarily seems so random and scattered, so fleeting and derelict, is brought into an exact unity that heightens the vibrancy of its sensuous presence. Such epiphanies have inspired this person to take the steps necessary to be one who produces things with that kind of power, steps that mean entering communities—first of fellow students, then of colleagues—formed around similar experiences and inspirations (1995, 9).

In design the world that ordinarily seems so random and scattered, so fleeting and derelict, is brought into an exact unity that heightens the vibrancy of its sensuous presence.

Donald Schön's descriptions of the kind of discourse that occurs in studio work capture the distinctiveness of the language of design. His examples of closely monitored sessions of instruction and critique reveal a kind of communication that may employ drawing as much as speaking, that often alternates between introducing geometrical forms and then reshaping them in response to particular features of a site. At various points there may be reference to precedents, drawing on a repertoire of design solutions that are the legacy of influential architects of the past (Schön 1983, ch. 3). It is not possible to reduce such discourse to a terminology or a method. The senses and the imagination are at work in it at every step; they are developed and integrated into the discourse over countless hours spent in the production of renderings and models. To know just what is being said when an architect speaks the language of design is difficult without having experienced something of that kind of training. Snodgrass and Coyne recognize these patterns as a matter of designing within the Gadamerian hermeneutical circle. In it, ideas, conditions, and precedents come into a dialogue to mutually inform one another and move the process forward along paths that could not have become apparent by any other means (2006, 45–8).

Accomplishment in the art of design, and fluency in its discourse, grant architects the irreplaceable expertise that they bring to projects, yet to their frequent dismay, Hubbard notes, the discourse of design cannot function as the common currency among the collaboration's participants (1995, 10). Schön and Hubbard both recommend that architects appropriate this fact; that they reflect on the powers and limitations of the world of discourse at the heart of their training. For Schön this involves demystifying the expertise of the professional and seeking to grasp how non-professionals hear professional discourse (Sirowy 2010, 196–201). For Hubbard, similarly, it is a matter of bringing the whole of one's values to the project, so that one relates to other participants not only from the perspective of the professional but through all of the other ways that people relate to the project (Hubbard 1995, 14–16, 166). Such recommendations begin to approximate the Gadamerian idea of the fusion of horizons within the context of shared practical endeavor. The "fusion" should not be taken to imply a compromise of the possibilities of design or a surrender of art to pragmatic considerations. On the contrary, the intertwining of perspectives and the exploration of assumptions can create the kind of trust and appreciation that grants the architect more permission rather than less. The hermeneutic approach should not be thought to be a way around the struggles of negotiation, the heated moments, and the close votes that are typical of collaborative undertakings in architecture. But these may be more productive for everyone if they occur *within* the effort to understand one another rather than instead of it.

The example of the rural studio

Beata Sirowy has drawn upon insights from both Gadamer and Schön (among a number of related theorists) in her study of the work of the late Samuel Mockbee and the work of the Rural Studio in the State of Alabama in the U.S. Conceived as an integration of architectural education at Auburn University and service to local communities, the Rural Studio involves students in designing houses for residents in Hale County, a region of great natural beauty and harsh rural poverty. The projects of the Rural Studio give clients their first experience of working with architects and give many of the students their first experience of life in one of the most disadvantaged regions of the country. Into this setting Mockbee introduced a process that had unusual expectations for everyone involved. He wanted to create homes that would not only serve basic needs but would embody the character and spirit of those who would dwell in the homes, along with the spirit of the region in which they dwelt.

To achieve this goal Mockbee conceived an organic process in which getting to know the members of the community would create a conversation that would shape the design. In the play of ongoing efforts to understand one another—the clients learning to respect the dedication of the students, the students learning to recognize the horizons in which the clients found meaning and purpose—the elements of design possibilities would organically emerge. These elements would draw on traditional styles of ordinary buildings in the region, and would respond to the strong light and heavy rains characteristic of the climate. But the designers did not resist innovation, and many of the buildings feature unusual elements—most obviously, the frequent use of "overstated" roofs that create places where residents can gather outdoors under shady eaves. Innovations resulted also from possibilities that are created by the heavy use of inexpensive, donated, and repurposed materials. The design imagination of students and professionals was put to work discovering new ways to draw out the aesthetic possibilities of materials created for sometimes very different purposes. Multiple designs worked out by students in the studio would be assessed, selected, reworked, and then taken to clients for further input. Thus clients were involved continually in the process (Sirowy 2010, 239–49).

The example of the first building created by the Rural Studio, the Bryant "Hay Bale" House, illustrates these qualities well. Alberta and Shepard Bryant were in their seventies and raising three grandchildren. Out of the conversation as to how they were living and what sort of spaces would be desirable, elements emerged such as the need for protection from the rain, the desire for each child

Rural Studio, Bryant "Hay Bale" House, Hale County

to have a room with a bed and a desk, and the value of a porch where people could gather and spend time. The use of hay bales covered with stucco was an unusual strategy, but one that proved well-suited to providing protection from the weather at low cost. Three small rooms were created in the back of the house, and are visible as three semi-circular forms on the rear elevation. A strikingly large overhanging roof made of translucent corrugated Plexiglas and supported by recycled beams and traditional-looking columns creates a large social space recalling the luxurious verandas of antebellum mansions. The house combines features of great simplicity with elements of imaginative extravagance. In articulating how its residents use their living space it expresses their relationship to a distinctive landscape and climate. In recalling traditional forms of the region while incorporating innovative materials, the architects bring tradition forward in modest, respectful, and sometimes eloquent ways (Sirowy 2010, 252–6).

> In recalling traditional forms of the region while incorporating innovative materials, the architects bring tradition forward in modest, respectful, and sometimes eloquent ways.

Sirowy notices many corollaries to Gadamerian ideas in Mockbee's inspired approach to socially engaged architecture. The approach is attuned to qualities of local life and the spirit of places. It draws upon traditions by bringing them forward and interpreting them in contemporary terms. It is committed to the kind of dialogue that brings out the questions, assumptions, habits, experiences, and stories that shape individual and community horizons. But at the same time it respects the insights and techniques that theory can introduce into the conversation. The process manifests an understanding that there is no such thing as a general fusion of horizons (Sirowy 2010, 257–60). To the extent that fusion occurs is does so in unique relationships that must be forged by the actual participants in the project. In the unfolding of the process many principles come into play that one is inclined to call universal—justice, human dignity, sustainability, professional integrity, honesty—but with the understanding that these can only be realized in the concrete as qualities of the relationships that the shared work of the project initiates. The universal cannot be taken for granted; it must speak again in the unique circumstances of the here and now.

Finding measure

The case of the Rural Studio illustrates a conviction that has been expressed by a number of writers regarding the process of architectural creation: that the challenge of opening up to the experiences and points of view of non-architects can be a means of liberating architectural creativity rather than constraining it (Sirowy 2010, 258–61; Till 2009). Bruce Lindsey, a colleague of Sam Mockbee whom Sirowy quotes, expresses the idea in these words:

> **The Rural Studio emphasized the fact that when you advocate for something outside your discipline—like people and communities—it opens up the possibilities of your own discipline. In other words, when you advocate beyond narrow technical, aesthetic, or professional interests, the opportunities for every aspect of what you do open up. The innovative, expressive, and spiritual aspects of architecture are no longer confined.**
>
> (Sirowy 2010, 261)

To take advantage of such opportunities requires finding measure—learning when to listen and when to reflect, when to introduce one's expertise and to experiment, when to be critical or to raise difficult questions, when to push the process, set deadlines, and just get things done. Ultimately the possibilities that open up are not merely possibilities for one's professional work, but also for integrating professional work into the whole of one's life and character.

Questions regarding this kind of measure—of balance and proportion, of occasions for thought and occasions for action—are among humankind's oldest. Gadamer's studies of the Greek thinkers that lived before Socrates find them learning from the balance of nature's forces a model for the order of the human psyche. A Greek phrase that has come down to us—*metron ariston*, "moderation is best"—can be easily interpreted as requiring us to avoid excesses of passion and arrogance, but it equally has in view a way of adding experiences and insights to life that further cultivate the passions and humble the self-aggrandizing ego with an appreciation of the greatness of the world. To develop in this kind of moderation is to develop in practical wisdom, and there is no activity of the human person that cannot benefit from that kind of growth of spirit.

CHAPTER 6

Architecture as a Way of Being

The most philosophically technical and daunting feature of Martin Heidegger's thought is its grounding in a philosophy of being, or "ontology," and yet it is the profundity of this grounding that has made Heidegger so fascinating to non-philosophers, including those in architectural fields. The proximity of Gadamer's thinking to that of Heidegger has much to do with their shared interest in a "hermeneutic ontology," thus there is value in considering in more detail the ontological dimension of Gadamer's philosophy. In the course of doing so one discovers important respects in which Gadamer's thought differs from that of Heidegger, and these points of divergence will be seen to produce corresponding differences in the two thinkers' relationship to architecture. Gadamer's work in ontology and language has also brought his philosophy into a relationship with postmodern thought, a movement that has had its own powerful influence on contemporary architecture.

Some puzzles about being

The question of the meaning of the word "being" would appear to be as old as philosophy itself. From the earliest records of Western intellectual history one can find thinkers wondering about what might be meant when one says that something "is." The word can be taken to be a simple copula, a means of tying predicates to subjects—as in "the stone is white"—yet the quality of being, or of existing, belongs independently to both the subject and the predicate: the stone *is*, whiteness *is*. From this insight we might want to say that being is itself a predicate, but if we do so we must admit that the quality of existing contained in this predicate is different from other kinds of qualities of things. To say the stone is white, or heavy, is to identify qualities that exclude other qualities. If the stone is white it is not black; if it is heavy it is not light. But to say that the stone "is" does not exclude anything; existence pertains to everything, including both the stone and any qualities that one could ascribe to it. Words like "white" or "heavy," furthermore, are abstract categories, but what is meant by "being" must be utterly concrete. Indeed being, many philosophers have said, is precisely what makes a thing concrete and actual rather than merely abstract and possible.

Further puzzles about being come to light when we ask whether we *know* being. On the one hand, there is reason to say that being is more known to us than anything else, for we ourselves exist and the sense of our own existence gives us the idea of being by which we know anything else to exist. But on the other hand, if being pertains to everything in the universe that exists, then it seems absurd to say that we know being, for we know only a tiny fraction of all that exists. Being would thus seem to be a mysterious known-unknown—intimately familiar yet at the same time unfathomably beyond our powers of comprehension.

Being would thus seem to be a mysterious known-unknown—intimately familiar yet at the same time unfathomably beyond our powers of comprehension.

Heideggerian thinking of being

Heidegger's thought came to be motivated by the conviction that much of the work that Western philosophy had done to address the question of being merely covered it over. Western metaphysics concerned itself with substances and qualities, becoming satisfied with a universe composed of discrete entities and their attributes instead of thinking deeply about the mysterious coming-to-be and passing-away of these things, and the uncanny question as to why there should be anything at all. Modern science became similarly satisfied with the study of "objects" and their practical manipulation. Time came to be viewed as the container for the world's objects rather than the enigmatic movement that defines things, at their core, as coming-to-pass.

Heidegger's ontology, then, aims to recover questions that have been latent, but only weakly pursued (according to him), in the whole of the Western tradition.

Heidegger struggled to fathom the mystery of being through the appearing of the phenomena of this world. To do so required getting past the common tendency to imagine the world as one of subjects confronting objects, or as objects impinging on the senses of subjects, for that assumption already moves away from the mystery—beginning, instead, with unexamined metaphysical assumptions about the reality of subjects and objects. Husserl's phenomenology appeared to Heidegger as a means of setting aside some of these assumptions, because it attempted to describe experience *before* distinguishing the subjective and the objective. But Heidegger felt that phenomenology, too, had to be radicalized in order to serve as a means of renewing the question of being (Heidegger 2010, 1–10). Phenomenology had to be made hermeneutical, in the sense that it had to reflect upon being with a clear awareness that being must be thought from *within* being, bringing to light from the very outset the complete interwovenness of thinking and being. Heidegger had to find some way of saying how human openness to the question of being constitutes being's openness to itself, yet without claiming, as Hegel seemed to do, to grasp the ultimate structure of being. To steer this course Heidegger drew on the existence-philosophy of Kierkegaard and the life-philosophy of Nietzsche to show how our deep intimacy with being is circumscribed by the finitude of our existence (Heidegger 2010, 32–7; Gadamer 1990b, 265–71; 1994). The appropriation of this finitude must shape every step of the meditation on the meaning of being.

It is out of these motivations that Heidegger's works present us with unusual formulations of classic philosophical problems. In *Being and Time*, for example, rather than producing a philosophy of human nature or human consciousness, Heidegger proposes a "hermeneutic of *Dasein*," wherein a literal reading of the word *Dasein* (which English editions normally leave untranslated) as "there-being" refers to human being purely in terms of its ontological openness. Any approach to the ontological event of openness that construes it as subjectivity, or as consciousness, or as the perception of objects is already beginning to give this ontological openness a secondary status. In later writings Heidegger moves even further from typical ways of posing the problem. Rather than speaking of "*Dasein* in its openness to being," he focuses more strictly on the openness *per se*—the "lighting" or "clearing" in being by which being is opened. The opening is "*Ereignis*," a word that suggests "event" but also "appropriation"—the disclosure of being, one might say, by which being takes possession, in some measure, of itself (Heidegger 1993b; 1993c; 1999).

Even in this brief indication of Heidegger's project, one can already see that for him language was as much a foe as a friend. Language institutionalizes all of the paths by which being has been forgotten in the pursuit of other concerns, yet language also retains the echoes of earlier wondering, permitting a kind of ontological meditation on words that retrieves possibilities of thought that are hidden within them. In the course of such meditations Heidegger became increasingly focused on poetic language, for poetry pushes language to its limits and seeks to reveal, through language, that which language has so far failed to say. Heidegger's own reflections on word etymologies, such as the sources of the word "*bauen*" (to build) in his essay on architecture, have their own poetic qualities that stretch the definition of philosophical inquiry beyond its usual bounds (Heidegger 1971, 145–61; Sharr 2007).

For Gadamer it is important to stress the religious motivation in Heidegger's philosophical program. It began as the work of a young Catholic theologian encountering Luther's re-centering of Christian life in the finitude of humanity's sinful state and the event of Christ's sacrifice. Heidegger's project retained characteristics of a spiritual journey as it undertook an analysis of the human way to be that yields affirmation and resolve regarding the "finite transcendence" of *Dasein*'s existence. It acquired qualities of mysticism in its seeking after "poetic dwelling" and qualities of monasticism in its devotion to the simple life of thinking in the remoteness of the rural Black Forest. Everything in Heidegger's thinking is dedicated, in one way or another, to the kind of spiritual revelation that turns the world of the ordinary into something utterly extraordinary and mysterious. But this is a decidedly world-immanent revelation, bringing meaning to worldly involvements in which we are always already caught up rather than seeking intimations of a world that lies beyond experience.

For Gadamer it is important to stress the religious motivation in Heidegger's philosophical program.

Hermeneutics and Heideggerian ontology

Gadamer never hesitated to acknowledge the brilliance and depth of Heidegger's thinking, and a large array of Gadamerian ideas exhibit the stamp of Heideggerian influence. Heidegger recovered, for Gadamer, the "*Ereignis* character" of the work of art that allowed Gadamer to connect artistic play with general patterns of human understanding; Heidegger gave Gadamer the idea that the hermeneutic phenomenon is an ontological phenomenon; Heidegger developed the conception of truth that is invoked in the title "*Truth and Method*"—a conception of truth as that which, coming out of the darkness of the undisclosed, is surrounded by that darkness, shaped by it, and prone to recede into it (Heidegger 1993a; Gadamer 1994, 91–3). The list of such Heideggerian influences could be expanded at length, yet they would never add up to a picture of Gadamer as a straightforward Heideggerian, even on matters of ontology (cf. Schmidt 1994).

Because Gadamer could find, at nearly every phase of Heidegger's development, something to appreciate in his thought, Gadamer's points of disagreement with Heidegger, though unambiguous, can fail to stand out for his readers. A very common point of resistance concerns the remoteness of Heidegger's terminology. The resistance can be seen already in Gadamer's use of the word "consciousness" rather than Heidegger's "*Dasein*." For Heidegger the new ontology requires terminological innovation; for Gadamer communication of the new philosophical insight calls for making an impact on the language that is already spoken and shared. Gadamer's resistance to Heidegger's language intensifies as that language becomes ever more idiosyncratic and rarified. Gadamer tells the story of listening to Heidegger express his own frustration with the linguistic contortions to which he was resorting in his late writings: "But this is Chinese!" Heidegger said. Gadamer comments: "And he was right. It was" (1992, 128).

Behind Gadamer's resistance to Heidegger's linguistic innovations is a disagreement regarding the polemical spirit that drives them. Heidegger's relationship to the history of philosophy always combined the intentions of "destruction" and "retrieval"—the motives of polemic and hermeneutic recovery. In this combination of aims Heidegger was habitually hasty, in Gadamer's view, in assuming that the way he attempted to think had not been attempted in the tradition and could not be thought through traditional language. This polemical bias of Heidegger's has a dual consequence: ideas from the tradition are rejected before they are fully understood, and what is "retrieved" is often a matter of Heidegger recognizing something of his own thought in the thinkers he interprets. As Heidegger railed against "forgetfulness" and the "oblivion" of being in Western thought, Gadamer was the one always pointing out that a moment of forgetfulness is always also an opportunity for retrieval (Gadamer 1986d).

Avoiding, for the most part, then, the development of his own unique terminology, Gadamer characterizes his philosophical project as Heideggerian in its efforts to find the *Ereignis*-dimension—the being-character as it comes to light—in many forms of human endeavor. It is what Gadamer is seeking when he not only speaks in philosophical terms of "temporality" and "historicity" but immerses himself in the study of history and historiography, taking a special interest in the history and theory of the hermeneutic tradition. In carrying the Heideggerian project forward in this way Gadamer also alters it. This combination of dedication to Heidegger with a reorientation of the ontological project is what Jürgen Habermas has in mind in saying that Gadamer "urbanizes the Heideggerian province" (Habermas 1983, 190). It is its movement away from Heideggerian insularity that allowed Habermas to see so much relevance, in Gadamer's hermeneutics, to the contemporary study of human society.

The ontology of building and place

The power and influence of Heidegger's writings on art and architecture has much to do with the fact that they are essays in ontology rather than being simply aesthetic or phenomenological. Heidegger's famous reversal of the notions of "building" and "dwelling," for example—wherein one thinks of dwelling less as something that happens within buildings than as a state of being in the world that illuminates how we should build—is a shift in awareness from beings to their larger ontological context and ground (Heidegger 1971, 145–9). By effectively raising architectural thinking to the level of ontological thinking, Heidegger has become a crucial figure in contemporary efforts to reconnect architecture with ultimate human concerns. His thinking has played a role, for example, in the interdisciplinary effort to recover the richly human connotations of the notion of "place" from its reduction to technologies for the manipulation of "space" (Casey 1993; 1997). It becomes key to Karsten Harries' efforts to steer a course for architectural ethos that is neither the aestheticism of the Baroque nor the moralism of moderns like Loos. Heidegger connects with primordial experiences of the immediacy of being that has eluded modernity altogether (Harries 1997, 160–62). Yet there is a serious question, in Harries' mind, as to whether Heidegger's archaic-sounding expression of the realm of symbolic meaning in terms of "earth, sky, mortals, and divinities" can really be made relevant to modernity. In a similar vein, Harries wonders what is to be made of Heidegger's treatment of the pre-modern Black Forest farmhouse as a paradigmatic example of fully realized architectural meaning (1997, 162). How would one translate such things into the modern context? How would one build with the same sense of place in very different landscapes and with very different techniques and materials?

By effectively raising architectural thinking to the level of ontological thinking, Heidegger has become a crucial figure in contemporary efforts to reconnect architecture with ultimate human concerns.

Christian Norberg-Schulz addresses this problem by extending the scope of Heideggerian architectural reflection, combining it with studies in the history of architecture aimed at bringing out the meaning-dimension in a great range of historical contexts (Norberg-Schulz 1975). Over a series of books he develops an expanded set of interpretive terms based on fundamental architectural dynamics, adding to Heidegger's dialectic of earth and sky, of mortals and immortals, relationships such as that of settlement to landscape, and of path to place (Norberg-Schulz 1979; 1985). Norberg-Schulz often begins close to Heidegger, explicating symbolic elements within vernacular forms of architecture, but moves on, then, to interpret major iconic works and the organization of whole towns and cities. To a degree, one can say that Norberg-Schulz, by this historical application of the Heideggerian perspective and this expansion of terminology to incorporate terms familiar to architects, moves Heideggerian thought on architecture in a Gadamerian direction. But there are complications involved in making this claim, for it is not always clear in what sense, and to what degree, Norberg-Schulz has appropriated the specifically ontological dimension of Heidegger's thought and to what extent Norberg-Schulz's approach functions as a phenomenology (rather than an ontology) of architecture, or even simply as a typology of architectural forms. The history of architecture in which Norberg-Schulz immerses himself, moreover, is a history seen *through* his typological scheme rather than by means of a hermeneutical study of the ideas that inform various ages of architectural thinking—as Harries, Leatherbarrow, and Pérez-Gómez have pursued. These are reasons for saying that the latter authors perhaps offer better examples for illustrating how a Gadamerian approach to architecture might make advances on a Heideggerian one.

Robert Mugerauer employs yet another strategy. Working from the hypothesis that Heidegger's thinking is inseparable from his language, Mugerauer goes further than perhaps any other writer in mining the details of that language for insights into the ontological meaning of home, homelessness, and homecoming (Mugerauer 2008). One may read his effort as an expansion of the Heideggerian notion of "dwelling on Earth" as it forms a pivotal dynamic in Heidegger's late ontology. Human being in its ordinary business among things is oblivious to the uncanniness (*Unheimlichkeit*) of existence, the state of being not-at-home (*unheimlich*) in the world because one is uniquely opened onto the question of being. To discover that uncanniness requires venturing into unfamiliar terrains of thought, though all the while seeking new ways to be at home in being. Mugerauer traces in meticulous detail how this 'venturing out' led Heidegger into the language of poets, how he came to blend poetic speech with philosophical terminologies, and how poetry and philosophy came together to yield Heidegger's highly original formulations of ontological insights. The particularities of the Black Forest landscape are important to these meditations in that they embodied, for Heidegger himself, the elemental qualities of "home" from which one ventures, toward which one journeys, and regarding which one discovers duties to protect and preserve.

Every Heidegger scholar who moves, in this fashion, toward the mystery of being in its most evocative Heideggerian expressions must wrestle with questions of particularity and generality. To what extent is one meant to dwell *with* Heidegger in his particular landscape, among the traditional architecture of Todtnauberg, and the cabin in which he composed so many of his captivating thoughts (cf. Sharr 2006)? To what extent are Heideggerian insights tied to the poets that he cites and the rarified terminology that his work develops? The point of Heidegger's writings (as Mugerauer and others make abundantly clear) is for every reader to rediscover his or her own experience of finite openness to being within the circumstances of his or her life. Meditation on being is meant to be transformative of one's own thinking, never to let Heidegger do the thinking for one. But this aim blends in often ambiguous ways with Heidegger's own struggles in the task, which he undertakes insistently in terms of his own particulars, and with a polemical agenda that often seems to abandon hope for the horizons that most readers would naturally bring to his texts.

Meditation on being is meant to be transformative of one's own thinking, never to let Heidegger do the thinking for one.

Intimately familiar with Heidegger's intentions and ambiguities, having made many visits to the *Hütte* in Todtnauberg, and having been present at many of the lectures that have influenced aestheticians and architectural theorists, in this area Gadamer is once again surprisingly distanced from the particularities of Heideggerian terminology and imagery. His writings and interviews express appreciation for the efforts of Heidegger scholars to be rigorous and exact, but he fears that the attempt to render Heidegger's language with utter faithfulness turns too easily into a kind of faith in Heidegger. He wanted scholars to emulate Heidegger's restlessness with formulations and to seek formulations that reflect their own circumstances and questions.

The ontology of language

Heidegger famously said that "language is the house of being." In this statement one can hear the echo of the Socrates' belief that language has much to teach the philosopher, and between these two—Socrates and Heidegger—one can see Gadamer finding the path to his own famous declaration: "Being that can be understood is language" (1990b, 474). By this statement Gadamer means that every philosophical reflection on being must be a hermeneutical venture into the ways in which being comes to articulation. As the medium for understanding and communication, language forms an ultimate horizon. Yet even this ultimate horizon, one must insist, is an open horizon. Language is always opened out on to an indeterminate potentiality of meanings, and is thus everywhere caught up in the mysteries of the known-unknown. For this reason Gadamer never means by "language" simply a system of signifiers. That would be a closed horizon. He always means also the existential activity of putting verbal meanings into action and formulating experiences into words. Language lives in the ongoing process of understanding and discovery that adapts and alters it.

The claim that "being that can be understood is language" is sometimes taken by readers of Gadamer to suggest that non-verbal forms of expression—such as architectural forms—must have a less-than-evident ontological status or must somehow fail to be comprehensible. But in fact Gadamer includes these forms of meaning in his broad notion of language or linguisticality (*Sprachlichkeit*). Speechless artistry comes to exist within a movement that departs from speech and heads towards speech—attempting, indeed, to embody what language fails to say, yet nonetheless in a manner that "speaks" to us. In this speechless "speaking" one may experience many of the suggestive gestures or indicators that function like inchoate speaking. As Gadamer explained in his notion of visual art's "stammering," the speechless work may be less devoid of the verbal than it is bursting with so much to say that it can articulate no one thing.

More fundamentally, one can say that the whole of Gadamer's philosophical endeavor is dedicated to the struggle to find the mysterious movement of being by which the particular and the incarnate *come to speech*, attain a level of communicability, and then, in turn, shape experience itself. It is this fascination of Gadamer's that leads him to focus, in the third part of *Truth and Method*, on the medieval thought of Thomas Aquinas on the subject of *verbum interius*, or the "inner word" which arose out of the efforts of Augustine and Aquinas to find analogies in human experience to the Christian mystery of the "word made flesh" of John's gospel. The distinction of "inner" and "outer" word, in this context, should not be taken as meaning "within the psyche" versus "out in the world," but rather the more primordial sense in which the intertwining of actuality and intelligibility makes its way into speech and (because of the inevitable incompleteness of verbal expression) draws one back into the richness of the experience (Gadamer 1990b, 418–28; Arthos 2009; Grondin 2003, 134–7; Risser 2007). By virtue of this sense of intertwining Gadamer's ontology of language stands, like Heidegger's, at the boundaries of speaking, with the poets and artists who mine the suggestions and ambiguities of words for their revelatory power. The horizon of language forms a limit to speaking, but also always an opening onto the not-yet-spoken.

The ontology of time

A definitive claim of Heidegger's ontology is that *Dasein*'s way of being is time, or temporality. When we think about time, our temptation is always to imagine ourselves a stable entity moving through time, or riding along the stream of time like a floating vessel. Heidegger's *Being and Time* seeks to shift this mentality, to say that we are not something that moves through time but something that exists, rather, as the movement itself. This means that the experience of presence (both in the sense of the temporal present and the sense of being present to the world) emerges only out of the constant movement of an anticipated future into a receding past. Presence is not a static state but is always coming-to-pass. We are never simply present to the world; our experience of presence results from the working of that which is not yet and that which no longer is—two unimaginably vast forms of absence.

Presence is not a static state but is always coming-to-pass.

To appropriate our being, therefore, is to appropriate our temporal finitude, realizing the great darkness, stretching before and after, in which a flickering light of experience, understanding, and truth has been lit. But in this realization there is an instance also of our transcendence, of our astounding ability to be aware of being, which makes us more than a thing that merely transpires in the dark movement of time. The moment of realization (*Augenblick*) is itself an intensification of the light, and there is a sense in which time, in such a moment, is gathered together in a disclosure of its significance (Heidegger 2010, 312–14; Sheehan 2001).

Heideggerian notions of time were thoroughly influential on Gadamer's thinking—most obviously in the way temporality is constantly in view in the development of Gadamer's ideas on horizons and historicity. However here, as in other areas, Gadamer emphasizes the continuity of Heideggerian insights with a whole history of philosophical thinking on time, rather than dwelling on Heidegger's departures from that history, and Gadamer looks for the moment of realization even in very common experiences (Gadamer 1970; 1972). The arresting experience of being moved by a work of art can have such a character. In the moment on the stairwell when the artistry of a building suddenly emerges from the background and shines forth, the activities and purposes that one had been engaged in cease to be merely a succession of events and suddenly seem to acquire a new importance—an "increase in being."

The example indicates the role of temporal ontology in Gadamer's relevance to architecture. One can see that his frustrations with an architectural aesthetics that is increasingly visual and instantaneous arises from his conviction that architecture is fundamentally a temporal art, that it reveals our way of being best when it unfolds for us in the kind of narrative manner that our lives do. It is only within such a temporal unfolding that the moments that gather time together—moments such as that of Gadamer's stairwell, or Holl's archetypal experiences, or Hubbard's design epiphanies—occur with the full force of their significance. The "decorum" of architecture, to put it another way, is not simply an aesthetic or experiential quality, but has its roots in the temporality of being.

The hermeneutics of suspicion

The phrase "hermeneutics of suspicion" was coined by Paul Ricoeur to categorize forms of interpretation that cast doubt on the evidence of consciousness, rationality, and the ego. The intellectual movements that Ricoeur indicates by the term share a source in Nietzsche's radical attack on the entire edifice of Western conceptual thought, in which he claimed that every abstraction attenuates the richness of life, draining it of its force, reducing it to desiccated "concept-mummies," which the practitioners of Western metaphysics then stuff into the great columbaria of theoretical systems. As these thinkers do so, Nietzsche claims—as they posit an imaginary realm of universal, transcendent, conceptually articulated metaphysical reality—they actually behave, in their lives, the same way all human beings behave: according to the desire for vitality, mastery, and creative control—in short, the will to power . The suspicion, then, is that all claims of conceptual rationality—the very core of the philosophical tradition—are naïve, or disingenuous, or a combination of both. For Nietzsche, the highly intellectualized moralism of Socratic philosophy is an early symptom of cultural decay, a falling-away from a belief in the dignity of the struggle of life that is embodied in the character-formation of the noble warrior and is raised to high art in Greek tragedy (Nietzsche 2006, 114–23, 456–85). Nietschean thought has decisive implications for the question of interpretation, says Gadamer:

> The "will to power" changes completely the idea of interpretation; it is no longer the manifest meaning of a statement of a text, but the text's and its interpreter's function in the preservation of life. The extension of power—that is the real meaning of our all-too-human insights and cognitions.
>
> (1984b, 58)

<u>The "will to power" changes completely the idea of interpretation; it is no longer the manifest meaning of a statement of a text, but the text's and its interpreter's function in the preservation of life.</u>

In the twentieth century Nietzsche's impact was felt everywhere in European and American culture, but perhaps most of all through the psychoanalytic theories of Freud, which cast light on the powerful life of unconscious human will that causes us to be always, at some level of our psyches, in conflict with the moral norms that our society imposes upon us. It was felt also in forms of Marxism that used Nietzschean suspicions to unmask capitalism's claims to universal economic rationality and to expose the will to domination behind that mask. Heidegger was profoundly influenced by Nietzsche's rejection of the rationalist foundations of Western metaphysics, which he saw continuing into the twentieth century as the metaphysical foundations for modern science and technology. But Heidegger took Nietzsche to be retaining, in certain respects, the kind of metaphysics of the will that one finds in one form in Kant and in a very different form in Schopenhauer.

A further radicalization of Nietzschean ideas took place among thinkers who became classified as "postmodern" and "post-structuralist." These include (among many others) Jacques Derrida, Michel Foucault, Jean-François Lyotard, Gilles Deleuze, and Jean Baudrillard. Particularly in French intellectual circles, the structural linguistics of Ferdinand de Saussure and Émile Benveniste, along with the structuralist anthropology of Claude Lévi-Strauss and the structuralist reworking of Freudian theory by Jacques Lacan, caused the matter at issue to be seen as broader than a conflict between conscious reason and vital will. Structuralism had revealed inevitable patterns of binary opposition in every kind of symbolization and sign-making that seemed to further undermine any autonomy that might be claimed for conscious rational thought. There are systems of signifiers at work that order language, consciousness, and meaning, but cannot be consciously mastered or subjected to control.

The body of work associated with structuralism and poststructuralism is far too vast to summarize here, and I would refer the reader, instead, to books in the *Thinkers for Architects* series that address them. But there is a special relevance to our subject of the deconstructive philosophy of Jacques Derrida because of a famous attempt to bring Derrida and Gadamer into dialogue in the 1980s that became a *succès de scandale* by virtue of its failure. At a conference in Paris in 1981 Gadamer presented a wide-ranging account of themes in his hermeneutic philosophy and Derrida responded with surprising terseness by simply posing three critical questions to him (Gadamer 1989; Derrida 1989a). A first question focused on Gadamer's claim that understanding requires a measure of good will. Does this not, Derrida wondered, reintroduce the very kind of metaphysics of the will that Heidegger had opposed? The second question referred to psychoanalysis and wondered if Gadamer was not unjustifiably privileging the workings of consciousness. The third concerned the idea of reaching agreement: is there not a kind of prejudice at work in the assumption that an encounter with another should reach agreement? Both the critical nature of these questions and the brevity of the attention that they grant Gadamer give the impression of two thinkers who are very far apart.

Many had hoped that a shared background in phenomenology and Heideggerian philosophy would give Gadamer and Derrida a common starting point for discussion, but a disagreement over Heidegger's use of Nietzsche makes for an almost immediate disparity. In Nietzsche's anti-conceptualism Heidegger found a voice that echoed his own struggles with the ossified metaphysical categories of Western thought, yet Nietzsche's philosophy of the will struck Heidegger as another kind of metaphysics. Derrida reads Heidegger as missing the transformation of philosophical language that Nietzsche initiates. Nietzsche's attack on conceptualism and his embrace of rhetorical, literary, and polemical forms of discourse creates an elusiveness in him that, in Derrida's view, constitutes a more radical break with metaphysics than Heidegger himself could appreciate—or had, himself, achieved. For even in its more poetic moments, Heidegger's thought is always seeking the ground of all that is, which is identical with the aims of the *logos*-philosophy of Plato and Aristotle (Derrida 1989b).

From this point of view, Gadamer would be even less radical and more logocentric than Heidegger, for Gadamer rejected many of Nietzsche's axioms concerning language, believing that Nietzsche had misconstrued the originating insights of the ancient philosophers and had asserted an arbitrary relationship between language and reality as a founding dogma rather than a plausible philosophical conclusion. The fundamental question that gave rise to the metaphysical tradition, it could be argued, is the question, "Why are things similar to one another? Why is there pattern?" To take the path that Nietzsche and Derrida take is to question the possibility, and even the desirability, of answers to that question. Abandoning hope (and/or interest) in any fruitful pursuit of intelligible realities, one will see no special status for the directed reasoning of consciousness over the metaphoric and metonymic associations of the unconscious. Heidegger and Gadamer, from this point of view, remain "metaphysicians" because, however much they have rejected metaphysical theories, they remain wedded to the question of "essence" that has inspired those theories.

To say that Derrida doubts the question as to the *why* of patterning is not to say that he is uninterested in articulated patterns. On the contrary, his immersion in structuralist linguistics is comparable in its scope and detail to Gadamer's immersion in classical philology. But Derrida's questions about the patterns are different. Even more than Gadamer, a Derridean would argue, Derrida is focused on the patterns that shape our speaking and acting *in spite of* our intentions. The structures of language have their way with us to a point far beyond our consciousness. But language does nothing so neatly as it has been imagined either in linguistics or the philosophical tradition. Every distinction, every articulated difference relies, for its distinctiveness, upon everything from which it is distinguished, thus retaining reference to its alternatives even as it differs from them—an ambiguity captured by Derrida in the ambiguous term "*différance*." The path that deconstruction takes is to follow various deferrals—associations, intimations, suggestions, sublimations—that texts imply even as they pursue an explicit goal of leaving these very things aside (Derrida 1982). Rejecting the gravity that locks the mind into the confines of discursive reasoning, the deconstructive project is often playful, with an artistry that Gadamer could certainly appreciate. Yet Gadamer could not but think that the project wrongly forces itself into a merely negative strategy yielding only a limited kind of insight:

> **you can only start with surprising transformations of false prior opinions, transformations tied to a moment in which a light of illumination suddenly flashes, yet disappears again once one again looks for the same figure in the text.**
>
> **(Gadamer, et al. 2001, 62)**

Even more than Gadamer, a Derridean would argue, Derrida is focused on the patterns that shape our speaking and acting *in spite* of our intentions.

Derrida's third question displays his suspicion of dialogue and the ostensible effort to "understand one another." Oral speech is subject to the illusion that the speaker is in control of language, that one can force language, by speaking it properly, into single, stable meanings. But in written texts the ways of language are better revealed in all of their ambiguity. It is in working with texts that deconstruction can best bring out the implications and connotations of which the author may have had no inkling. Derrida suspects that in the professed aim of understanding through dialogue (which forms the very heart of Gadamer's hermeneutics) a desire to assimilate the otherness of one's interlocutor. To "come to agreement" always means giving up something of one's distinctiveness, and so the aim of reaching agreement inevitably masks the desire to get the other to yield something of themselves. Derrida sees a greater deference in respecting disagreement, choosing a relationship that is not conditional on mutual understanding, but rather deliberately preserves the otherness of the other. In this light one can see why it might be important to Derrida *not* to engage in dialogue with Gadamer. Something of the aims and assumptions of deconstruction might be surrendered in that very act (Simon 1989; Madison 1989).

When we cast the positions of Gadamer and Derrida in this kind of opposition, we must eventually come to the question as to what, in practical terms, is at stake in the debate. Here I would say that the two could agree that it is a matter of seeking a form of human relations that is superior to manipulation and violence. For Gadamer, this has always been philosophy's worthiest goal. Hermeneutics makes at least a small contribution to that goal by illuminating some of the complexities that are involved in seeking understanding and reaching reasonable agreements. Gadamer hears in the potentialities of language an openness that can breach the differences that engender hostilities. Derrida takes quite the opposite view. Hostilities cannot end as long as the goal is a kind of linguistic assimilation, for that goal itself is biased in favor of homogeneity, as is evidenced historically in the struggles of the West to assimilate and eliminate local cultures all across the globe. But here, Gadamer would say, Derrida has underestimated the ability of language to create something that is neither assimilation nor opposition, but forms a third kind of achievement; and philosophical hermeneutics must question the sudden optimism that the philosopher of suspicion exhibits if he or she believes that the very otherness that for most of human history has been a basis for making enemies should now form the path to peace.

All of these observations regarding Gadamer and Derrida take the easy route of putting the matter in terms of a stark contrast of views. More complex is the task that a number of scholars have undertaken of determining what sort of accommodation or fruitful relation could be made between these two differing points of view. James Risser, for example, has repeatedly drawn attention to the fact that in the background of Gadamerian moments of recognition and experiences of truth is an undisclosed infinity of potential meanings that can never, because of our finite nature, come fully to light. There is, other words, an irresolvable quality to Gadamer's notions of meaning and truth that may put him closer to Derrida than one might expect (Risser 1997, 128–38, 163–8; 2000). Richard Bernstein has re-imagined the Gadamer–Derrida encounter as a series of critiques and responses that yields a productive tension:

> **On the one hand, Gadamer helps us to see that the kind of negotiation that Derrida takes to be so fundamental for responsible decision and actions is a form of *phronesis*. But on the other hand, Derrida brings out complexities and risks of *phronesis*; he warns against reducing it to technical calculation or simply following a universal rule. In his language, there is always a gap— an abyss—which we have to face in making responsible decisions, whether they are ethical or political decisions.**
>
> **(Bernstein 2008, 597–8)**

Adrian Snodgrass and Richard Coyne could be characterized as seeking an analogous accommodation of Gadamer and Derrida in the area of design education. They find, in Gadamer, an interpretive framework that can help to re-integrate the increasingly fragmented structure of architectural education, and they see his approach to architecture's history as a way to overcome the "design amnesia" that modernism has wrought, yet without lapsing into a mere conservatism (Snodgrass and Coyne 2006, 131–46). But because the task of carrying forward an architectural tradition in contemporary circumstances requires constant openness to what is new and other, there is a value, too, in Derrida's radical understanding of otherness and the necessary disruption of expectations that it imposes. In characterizing the influence of postmodernism in general and Derrida in particular upon the practice of architecture, Coyne rightly cautions us not to leap immediately to questions of form and style. The postmodernist architectural break with characteristic qualities of the International Style did not, I would argue, depart from modernism *per se* so much as it indulged in the same freedom of exploration that had always been characteristic of modernism in painting and sculpture. The irony and playfulness in many of the exaggerated quotations of historical architectural forms had a connection to postmodern theory, to be sure, but in a manner that often seemed to reduce postmodern theory to a style or a new form of aestheticism. Early deconstructive architects took the idea of "deconstruction" rather literally, creating buildings that seemed to be falling down, or imploding, or exploding in places. The crudeness of this appropriation of deconstruction, it might be fair to say, yielded something of the opposite of the humility and openness that Derrida sought. Indeed, it provoked architect Moshe Safdie's well-known denunciation of the architectural *hubris* that makes the users of buildings obey the dictates of a literally imagined deconstructivism, subjecting them, all the while, to a glorification of violence through destructive imagery (Goldberg 2009, 10–11).

Early deconstructive architects took the idea of "deconstruction" rather literally, creating buildings that seemed to be falling down, or imploding, or exploding in places.

Such transformations of Derridean ideas into building forms, in Coyne's view, miss what is perhaps the most important relevance of Derrida's thought to the field, which is a theoretical contribution with particular importance for education. Derrida's questions engage the whole of an intellectual tradition, and hence share theoretical space with the Gadamerian hermeneutic project, but into this space deconstruction brings a form of radical questioning that demands an openness beyond what even Gadamer had anticipated, and it is to this kind of radical openness that design must regularly aspire (Snodgrass and Coyne 2006, 95–107: Coyne 2011).

Transcendence and mystery

The history of philosophy reveals that the philosophical question of being has long been intertwined with the question of divine transcendence, or the being of God. Gadamer's thought shows, at regular points, an acute understanding and appreciation of this connection, and yet far less than one might expect from one so open to the life of religious traditions, Gadamer assiduously avoids producing anything like a religious philosophy or a philosophy of religion. For him the boundaries of human finitude render human philosophical categories utterly inadequate to the task of speculating on the divine. For that, Gadamer was fond of saying, "one would have to be a god." In this stance of Gadamer's, Frederick Lawrence has noted a central conviction of Kantianism that persists: to speculate beyond experience is to be embroiled in irresolvable antinomies (1990). Gadamer's background in Lutheranism, with its emphasis on the gulf by which sin divides the human and divine realms, is also sometimes mentioned as a source of his doubts about theology, and to this one should add that he was not himself a believer, though he spoke of frequently imagining how nice it would be to be one (Grondin 2003a, 335–6).

But the philosophical core of Gadamer's philosophical position on religion, I would say, is in the humble role that he sees for hermeneutics. As hermeneutics interprets traditions, so Gadamer's relationship to religious traditions is that of an interpreter. As his own hermeneutics must insist, one cannot interpret such traditions without a profound openness to their horizon of belief. I would say that Gadamer was not only aware of this hermeneutical quality in his openness to faith traditions, but was proud of it. It was another point upon which he could contrast himself with Heidegger, whom he frequently said was dedicated, throughout his philosophical career, to a search for God. The claim is surprising, given that Heidegger turned away from his early Catholicism and the Lutheran ideas that had informed his readings of Christian scriptures early on. It was Heidegger, after all, who later called Christian philosophy "a round square and a misunderstanding," (2000, 8) and who spoke of the "being" with which his philosophy was concerned as other than a supreme being. But it was never wrong, in Gadamer's view, to hear in "that which grants the being of beings" an entirely new way of approaching the question of the divine, a way that would resist the distortions that are introduced when one begins from the assumption of God as "a being" (Gadamer 1994, chs. 14–15). Yet in this fact there arises a sense that Heidegger can be thought to be *competing* with religious traditions for a more authentic notion of God, and that this might limit his ability to hear those traditions in their own voice. Gadamer's sense of himself as an interpreter carefully avoids any such competition.

When Gadamer turned to the theme of transcendence he would often do so by looking to the Greeks. He was always fascinated by the particularities of the bold challenge that Socrates and Plato posed to the religious assumptions of their contemporaries. In the dialogue *Euthyphro*, for example, a stubborn young priest easily defines religious piety as obedience to the wishes of the gods. But Socrates wonders what obedience can mean if by "the gods" we mean the Homeric characters who are always arguing and waging battles with one another. One can see in Socrates' question here the critical audacity of philosophy in the face of tradition, to wonder whether the divine might be something more than the rather irrational creatures found in the works of the poets. Implicit in Socrates' query, too, is an ethical challenge: are gods such as these really worthy of worship? But what Socrates offers in place of the Homeric theology is not an alternative religion but rather the restless wondering and learned ignorance of the philosopher. This is the posture that seems to Gadamer to be the proper one for contemporary philosophers as well. It does not oppose religious speaking, as the intellectual heirs of the Enlightenment do, yet it does not speak religiously. It does not offer a philosophical equivalent to religious beliefs, but it offers some possibilities for their mediation in thought. Indeed, at the end of his life, this is what Gadamer saw as a most urgent task: not that the great religious cultures of the world be overcome, but that they do better at understanding one another (2004, 140–41).

<u>At the end of his life, this is what Gadamer saw as a most urgent task: not that the great religious cultures of the world be overcome, but that they do better at understanding one another.</u>

With relatively little alteration, such an active openness to religious experience may be adapted by those with religious and theological interests. This is abundantly demonstrated by the great variety of essays in Andrzej Wierciński's collection *Between the Human and the Divine* (2002a); it is seen in the efforts of Arthos (2009), Lawrence (2000; 2009), and Wierciński (2002b) to take up Gadamer's analysis of inner word and Trinitarian theology in ways that move beyond the argument of *Truth and Method*; it is exemplified by Walter Lammi's association of Gadamer with the tradition of negative theology (2008); and in architectural studies it contributes to the interpretations by Lindsay Jones and Thomas Barrie of sacred meaning in architecture. But one would be hard pressed to adapt Gadamer in ways that would develop the full-fledged Christian theologies of the sort that Rudolf Bultmann and Karl Rahner developed out of Heidegger's ontology (Macquarrie 1955; Sheehan 1987). Gadamer worked for years with Bultmann, but was never drawn to his "demythologized" Heideggerian reading of the Gospels. The idea that Bultmann's use of Heidegger permitted a penetration to the core of the Christian message did not seem to stand up in the face of the texts (Gadamer 1984a, 11).

In this connection one can see Gadamer once again resisting any tendency to make hermeneutic ontology a philosophy that stands at the end of history as the final fruit of its progress. Hermeneutics, for him, is not an alternative language for the expression of religious, or even philosophical, truth; it is always a path *into* traditions of every sort. Gadamer grants that his reticence to develop something that looks like his own philosophy of ultimacy would seem to make him a philosopher of *Dasein* who is more concerned with the "*Da*" than with "*Sein*" (Gadamer 2004, 130). But an undeniably defining feature of this *Da* is the uncanny manner in which human consciousness fails to fit into the ordinary business of the universe, standing out, always, by virtue of its awareness and its wonder. That wonder is at the heart of all ontology, of all philosophy, and indeed, of every kind of religious seeking. Religion poses

> an unavoidable question for us—a hope perhaps, or rather, a task that unites us all in our mutual understanding. This ultimate ethical task cannot be separated from the one task of questioning and understanding our own existence.
>
> (2004, 143)

CONCLUSION

The Hermeneutically Minded Architect

The account of Gadamer's philosophy that I have presented in this volume describes not one kind of relevance of his philosophy to architecture, but several. I have organized these forms of relevance into separate chapters so as to emphasize the differences among the various kinds of value that an architect might find in hermeneutics. Yet because these different guises of hermeneutics embody a single hermeneutic pattern, they can also help to bring a sense of unity to the various situations and activities with which architects find themselves engaged. By way of conclusion to my discussion, then, it may be helpful to recall briefly the various forms of relevance that hermeneutics has for architecture, to indicate the unity that underlies them, and to close with a description of qualities that might characterize a hermeneutically minded architect.

We began with the idea of play in art. Gadamer insists that play is a defining quality of art, and that this quality is liberated when the artist hands himself or herself over to the inherent dynamics of play. By stressing, at the same time, the importance of play—its grounding presence in the most serious of art works and its affinities with many forms of truth-seeking—Gadamer's thought makes a strong case for the cultural significance of architectural creativity and meaning. For these reasons, an architect informed by the hermeneutic conception of play should feel creatively supported and, one hopes, emboldened in that creativity.

We moved from the topic of play to that of history, with a specific focus on the question as to whether architecture today should embrace its past or leave it behind. Gadamer's hermeneutics, by emphasizing the inevitable, but often underestimated, influence of the past upon the present, shows that the modernist dilemma of the present versus the past is largely a false dilemma.

The hermeneutic realization is that one can never simply repeat the past nor can one ever simply step out of history. Coming to understand this idea should encourage architects to think of what they do as a creative dialogue with architecture's history—an ongoing critical interpretation of the past. They may come to see the meaning of present and future architecture as something born of the dialogue with that which has gone before, but which also continues to resonate in contemporary life.

Architecture's role in the ongoing histories of cultures, we saw, grant it special affinities with the humanities. Architecture is imperiled, from the Gadamerian point of view, to the extent that it falls away from the humanities and comes to imagine itself to be largely a set of technical solutions to functional problems. Because it has an affinity to the humanities architecture is threatened to the extent that the humanities themselves are threatened—when professional organizations and institutions of higher learning begin to think that their primary mission, and the only way to compete, is to pour resources into the sciences and technical fields while allowing the liberal arts to languish.

Gadamer's characterization of hermeneutics as practical philosophy, we saw, helps to show that "being practical" need not mean simply being superficial or surrendering vision to circumstances. An awareness of the role of horizons and an openness to qualities of horizons can deepen the "listening" and negotiation that must be successful if an architectural project is to reach completion. So far from functioning as an obstacle to the realization of architectural ends, the qualities of understanding and community that are born of an earnest and thoughtful process are of value in and of themselves.

The fact that Gadamer's hermeneutics is conceived as a philosophy of being, an ontology, can help to remind the hermeneutically minded architect that architecture, like philosophy, has the potential to orient itself to the most profound ways of wondering about the mystery of existence. For all of its playfulness and immediate practicality, for all of the restrictive circumstances, the mixed motives of clients, and the public diffidence in the face of which architecture must bring about its creations, architecture always harbors the latent power to turn the mind and the spirit to ultimate things.

What unifies these different forms of hermeneutic insight—in play, historical understanding, humanistic learning, practical interaction, and ontology—is the unity of the hermeneutic phenomenon itself. In every one of the situations one acts purposively—by playing, inquiring, or interpreting—so as to have some effect upon a shared world; yet in so doing one finds oneself caught up in forms of meaning that are not of one's own making. The very desire to have an effect, the very structure of one's purposes, is itself something originating in the larger world of meaning in which one's consciousness is not only a participant, but a consequence, an effect. To be hermeneutically minded is, above all, to be sensitized to the working of this pattern throughout one's life and work. It is to look for the pattern, to find situations where an appreciation of the pattern might help make the difference between understanding and misunderstanding.

What unifies these different forms of hermeneutic insight—in play, historical understanding, humanistic learning, practical interaction, and ontology—is the unity of the hermeneutic phenomenon itself.

So what is it, finally, that distinguishes a hermeneutically minded architect? It is, one might say, a kind of sensitivity—to where play might be appropriate, to where clues to solving problems may be lurking in language and historical sources, to the moments where listening might be more appropriate than speaking, the moments where one must question one's own assumptions before questioning those of others. It involves knowledge, though one couldn't call an architect hermeneutically minded simply on the basis of a body of knowledge. Neither is it exactly a matter of experience or skill, though again, these have their role to play. It is a manner of attending, of exploring, of recognizing, anticipating, inquiring, playing with possibilities.

The qualities of the hermeneutically minded architect, one might say, intentionally recall the Socratic idea of wisdom and the circuitous route that Socrates believed one must take in its pursuit. Wisdom is not exactly a kind of knowledge, or skill, or a particular type of character, yet somehow it adds to these and completes them. Socrates was thought to be wise, but he insisted that any wisdom he possessed consisted chiefly of an awareness of his ignorance. This awareness engendered a restlessness that ran so deep as to make him a permanent seeker after greater understanding, a search pursued with the intensity of a divinely mandated mission.

Wisdom is as desirable as it is elusive. Everyone seeks wise leaders of state and wise physicians. If it sounds a little odd, a bit of an overstatement, to speak of a "wise architect," still, when one considers all of the challenges that architects must face, all of the differing decision situations that they must work through in serving the ambitions of art, the legacy of tradition, the limitations and opportunities of technologies, the expectations of stakeholders—surely it can seem as if it is not just wisdom that is needed, but the wisdom of Solomon, to steer a proper course through all of the risks and hazards, and to reach a worthwhile end.

For Further Reading

A brief account of a major thinker's work, such as I have attempted here, can aspire to be little more than a key that unlocks doors. One of these doors opens out on Gadamer's texts themselves. A reader who has followed my presentation of some of the central ideas of *Truth and Method* may be able, now, to turn directly to that book and see the thread of its argument even as it follows many twists and turns of ideas that have unfolded across more than two millennia of philosophy's history. But other readers may wish to stay with shorter presentations, and here I would point to the value of certain articles. Gadamer himself thought that, in some ways, "On the Circle of Understanding" (1988) articulated the outline of his hermeneutics better than *Truth and Method* does. The first two essays in *Philosophical Hermeneutics*—"The Universality of the Hermeneutical Problem" and "On the Scope and Function of Hermeneutical Reflection" (1976)—remain valuable statements of Gadamer's thought, particularly for those who are interested in the social relevance of hermeneutics. The essays on art in *The Relevance of the Beautiful* (1986c) also stand on their own—though, as I have argued, they are enriched when they are related to the themes of *Truth and Method*. (Readers who follow my citations to *Truth and Method* should note that I use the larger-format second revised edition rather than the smaller-format "Continuum Impacts" edition, which has slightly different pagination.)

Gadamer's own overviews of his career, such as the essay that opens *The Gadamer Reader* (2007) or the "Reflections on My Philosophical Journey" in *The Philosophy of Hans-Georg Gadamer* (Hahn 1997), can serve as relatively accessible ways to trace the course of Gadamer's thinking. This latter book also contains Richard Palmer's thorough bibliography of primary sources. For those who are especially interested in the biographical details, Gadamer's

Philosophical Apprenticeships (1985) and Jean Grondin's *Hans-Georg Gadamer: A Biography* (2003a) are both very readable.

Other readers may wish to seek the aid of further secondary sources, of which there are too many good choices to name. There are other short accounts of Gadamer's thought, such as those of Grondin (2003b) and Lawn (2006). There are collections of essays, including Dostal's (2002) and Hahn's (1997), that seek to cover a range of Gadamer's themes and interests. There are sources such as those by Grondin (1997), Mueller-Vollmer (1998), Ormiston and Schrift (1990), Palmer (1969), and Schmidt (2007) that fill out the picture of Gadamer's place in the history of hermeneutics and contemporary philosophical debates. And there are books such as Risser's (1997) and Weinsheimer's (1985) that provide close readings of *Truth and Method*.

For those who want to see Gadamer applied in an architectural context, the most explicit and thorough example is Snodgrass and Coyne's *Interpretation in Architecture* (2006). This book is focused on particular questions of education in design, but it also touches on a broad range of interpretive issues, including the question of deconstruction and the interpretation of non-Western architecture. Vesely's *Architecture in the Age of Divided Representation* (2004), though not as explicit on Gadamer, is the product of many years of bringing Gadamerian ideas into its author's own teaching, teaching that has influenced a number of the other authors that I have discussed. I hope that my discussions of other authors who draw, in one way or another, on Gadamer will have given an indication of how a reader's interests in particular topics in hermeneutics could be pursued.

Bibliography

Aristotle (1999) *Nicomachean Ethics*, trans. by T. Irwin, Indianapolis, IN: Hackett Publishing Co.

Arthos, J. (2009) *The Inner Word in Gadamer's Hermeneutics*, Notre Dame, IN: University of Notre Dame Press.

Augustine (1958) *On Christian Doctrine*, trans. by D. W. Robertson, Jr., New York: Macmillan Publishing Co.

Bacon, F. (1960) *The New Organon*, ed. by F. H. Anderson, New York: Macmillan Publishing Co.

Barrie, T. (2010) *The Sacred In-Between: The Mediating Roles of Architecture*, London/ New York: Routledge.

Bernstein, R. J. (2008) "The Conversation that Never Happened (Gadamer/ Derrida)," *The Review of Metaphysics*, 61:3, 577–603.

Brogan, W. (2008) "Figuring and Disfiguring Socrates: A Gadamerian Reflection on the Relationship of Text and Image in Plato's Philosophy," *Philosophy Today*, 52: supplement, 144–50.

Casey, E. S. (1993) *Getting Back into Place: Toward a Renewed Understanding of the Place-World*, Bloomington, IN: Indiana University Press.

—— (1997) *The Fate of Place: A Philosophical History*, Berkeley, CA: University of California Press.

Coyne, R. (2011) *Derrida for Architects*, London/New York: Routledge.

Davey, N. (2008) "Hermeneutical Application: A Dialogical Approach to the Art/Theory Question," *Internationales Jahrbuch für Hermeneutik*, 7: 93–107.

Derrida, J. (1982) "Différance," in *Margins of Philosophy*, Chicago, IL: University of Chicago Press, 3–27.

—— (1989a) "Three Questions to Hans-Georg Gadamer," trans. by D. Michelfelder and R. Palmer, in *Dialogue and Deconstruction: The Gadamer-Derrida Encounter*, ed. by D. P. Michelfelder and R. Palmer, Albany, NY: SUNY Press, 52–4.

—— (1989b) "Interpreting Signatures (Nietzsche/Heidegger): Two Questions," trans. by D. Michelfelder and R. Palmer, in *Dialogue and Deconstruction: The Gadamer–Derrida Encounter*, ed. by D. P. Michelfelder and R. Palmer, Albany, NY: SUNY Press, 58–71.

Descartes, R. (1993) *Discourse on Method and Meditations on First Philosophy*, trans. by D. A. Cress. Indianapolis, IN: Hackett Publishing Co.

Dostal, R., ed. (2002) *The Cambridge Companion to Gadamer*, Cambridge: Cambridge University Press.

Figal, G. (2010) *Objectivity: The Hermeneutical and Philosophy*, trans. by T. D. George, Albany, NY: SUNY Press.

Forester, J. (1993) *Critical Theory, Public Policy, and Planning Practice*, Albany, NY: SUNY Press.

—— (1999) *The Deliberative Practitioner: Encouraging Participatory Planning Process*, Cambridge, MA: MIT Press.

Fulford, R. (1992) "When Jane Jacobs Took on the World," *New York Times*, Feb. 16.

Gadamer, H.-G. (1985–95) *Gesammelte Werke*, 10 vols., Tübingen: J. C. B. Mohr.

—— (1970) "Concerning Empty and Fulfilled Time," *Southern Journal of Philosophy*, 8:4, 341–53.

—— (1972) "The Continuity of History and the Existential Moment," *Philosophy Today* 16:3–4, 230–40.

—— (1976) *Philosophical Hermeneutics*, trans. by D. Linge, Berkeley, CA: University of California Press.

—— (1980) *Dialogue and Dialectic: Eight Hermeneutical Studies of Plato*, trans. by P. C. Smith, New Haven/London: Yale University Press.

—— (1984a) "Articulating Transcendence," in *The Beginning and the Beyond: Papers from the Gadamer and Voegelin Conferences*," ed. by F. G. Lawrence, Chico, CA: Scholars Press, 1–12.

—— (1984b) "The Hermeneutics of Suspicion," in *Hermeneutics: Questions and Prospects*, ed. by G. Shapiro and A. Sica, Amherst, MA: University of Massachusetts Press, 54–65.

—— (1984c) *Reason in the Age of Science*, trans. by F. G. Lawrence, Cambridge, MA: MIT Press.

—— (1985) *Philosophical Apprenticeships*, trans. by R. R. Sullivan, Cambridge, MA: MIT Press.

—— (1986a) *Hermeneutik II: Wahrheit und Methode, Gesammelte Werke, Bd. 2*, Tübingen: J. C. B. Mohr.

—— (1986b) *The Idea of the Good in Platonic–Aristotelian Philosophy*, trans. by P. C. Smith, New Haven/London: Yale University Press.

—— (1986c) *The Relevance of the Beautiful and Other Essays*, trans. by N. Walker, ed. by R. Bernasconi, Cambridge: Cambridge University Press.

—— (1986d) "Hans-Georg Gadamer: Storie Parallele," *Domus* 671: 17–24.

—— (1988) "On the Circle of Understanding," in *Hermeneutics versus Science? Three German Views*, ed. by J.M. Connolly and T. Keutner, Notre Dame, IN: University of Notre Dame Press, 68–78.

—— (1989) "Text and Interpretation," trans. by D. J. Schmidt and R. Palmer, in *Dialogue and Deconstruction: The Gadamer–Derrida Encounter*, ed. by D. P. Michelfelder and R. Palmer, Albany, NY: SUNY Press, 21–51.

—— (1990a) "Reply to My Critics," in *The Hermeneutic Tradition: From Ast to Ricoeur*, ed. by G. L. Ormiston and A. D. Schrift, Albany, NY: SUNY Press, 273–97.

—— (1990b) *Truth and Method*, Second Revised Edition, trans. revised by J. Weinsheimer and D. G. Marshall, London/New York: Continuum. [Note: this is not the "Continuum Impacts" edition, which has slightly different pagination.]

—— (1992) *Hans-Georg Gadamer on Education, Poetry, and History: Applied Hermeneutics*, trans. by L. Schmidt and M. Reuss, Albany, NY: SUNY Press.

—— (1994) *Heidegger's Ways*, trans. by J. W. Stanley, Albany, NY: SUNY Press.

—— (1997) "Reflections on My Philosophical Journey," in *The Philosophy of Hans-Georg Gadamer*, ed. by L. E. Hahn, Chicago/La Salle, IL: Open Court, 3–63.

—— (1998a) *Praise of Theory*, trans. by C. Dawson, New Haven/London: Yale University Press.

—— (1998b) *The Beginning of Philosophy*, trans. by R. Coltman, London/New York: Continuum.

—— (2002) *The Beginning of Knowledge*, trans. by R. Coltman, London/New York: Continuum.

—— (2004) *A Century of Philosophy: Hans-Georg Gadamer in Conversation with Riccardo Dottori*, trans. by R. Coltman and S. Koepke, London/New York: Continuum.

—— (2006) "Architektur als 'Zuwachs an Sein': Hans-Georg Gadamer im Gespräch mit Catherine Hürzeler," in *Beyond Metropolis: Eine Auseinandersetzung mit der verstädterten Landschaft*, Zürich: Verlag Niggli.

—— (2007) *The Gadamer Reader: A Bouquet of the Later Writings*, ed. by R. Palmer, Evanston, IL: Northwestern University Press.

Gadamer, H.-G., Dutt, C., and Palmer, R. E. (2001) *Gadamer in Conversation*, trans. by R. Palmer, New Haven/London: Yale University Press.

Goldberg, P. (2009) *Moshe Safdie (Millenium) Vol. I*, Victoria, Australia: Images Publishing Group, Mulgrave.

Grondin, J. (1997) *Introduction to Philosophical Hermeneutics*, trans. by J. Weinsheimer, New Haven/London: Yale University Press.

—— (2003a) *Hans-Georg Gadamer: A Biography*, trans. by J. Weinsheimer, New Haven/London: Yale University Press.

—— (2003b) *The Philosophy of Gadamer*, trans. by K. Plant, Montreal and Kingston: McGill-Queen's University Press.

Habermas, J. (1983) *Philosophical–Political Profiles*, trans. by F. G. Lawrence, Cambridge, MA: MIT Press.

—— (1988) *On the Logic of the Social Sciences*, trans. by S. W. Nicholsen and J. A. Stark, Cambridge, MA: MIT Press.

—— (1990a) "The Hermeneutic Claim to Universality," in *The Hermeneutic Tradition: From Ast to Ricoeur*, ed. by G. L. Ormiston and A. D. Schrift, Albany, NY: SUNY Press, 245–72.

—— (1990b) *Moral Consciousness and Communicative Action*, trans. by C. Lenhardt and S. W. Nicholson, Cambridge, MA: MIT Press.

—— (1990c) "A Review of Gadamer's *Truth and Method*," in *The Hermeneutic Tradition: From Ast to Ricoeur*, ed. by G. L. Ormiston and A. D. Schrift, Albany, NY: SUNY Press, 213–44.

Hahn, L., ed. (1997) *The Philosophy of Hans-Georg Gadamer*, Chicago: Open Court.

Harries, K. (1997) *The Ethical Function of Architecture*, Cambridge, MA: MIT Press.

Heidegger, M. (1971) *Poetry, Language, Thought*, trans. by A. Hofstadter, New York: Harper & Row.
—— (1993a) "On the Essence of Truth" in *Basic Writings*, 2nd ed., ed. by D. F. Krell, New York: Harper & Row, 111–38.
—— (1993b) "Letter on Humanism" in *Basic Writings*, 2nd ed., ed. by D. F. Krell, New York: Harper & Row, 213–65.
—— (1993c) "The End of Philosophy and the Task of Thinking," in *Basic Writings*, 2nd ed., ed. by D. F. Krell, New York: Harper & Row, 427–49.
—— (1999) *Contributions to Philosophy (From Enowning)*, trans. by P. Emad and K. Maly, Bloomington, IN: Indiana University Press.
—— (2000) *Introduction to Metaphysics*, trans. by G. Fried and R. Polt, New Haven/London: Yale University Press.
—— (2010) *Being and Time*, trans. by J. Stambaugh, rev. by D. Schmidt, Albany, NY: SUNY Press.
Herzog, J. (2001) "Thinking of Gadamer's Floor," in C. C. Davidson, ed., *Anything*, Cambridge, MA: MIT Press.
Holl, S. (1999) *The Chapel of St. Ignatius*, Princeton, NJ: Princeton Architectural Press.
—— (2009) *Urbanisms: Working with Doubt*, Princeton, NJ: Princeton Architectural Press.
Holl, S., Pallasmaa, J., and Pérez-Gómez, A. (2006) *Questions of Perception: Phenomenology of Architecture*, San Francisco: William Stout Publishers.
Hollinger, R., ed. (1985) *Hermeneutics and Praxis*, Notre Dame, IN: University of Notre Dame Press.
How, A. (1995) *The Habermas–Gadamer Debate and the Nature of the Social: Back to Bedrock*, Aldershot: Avebury.
Hubbard, B., Jr. (1995) *A Theory for Practice: Architecture in Three Discourses*, Cambridge, MA: MIT Press.
Huizinga, J. (1950) *Homo Ludens: A Study of the Play-Element in Culture*, Boston: Beacon Press.
Innes, J. E. (1996) "Planning Through Consensus Building: A New View of the Comprehensive Planning Ideal," *Journal of the American Planning Association*, 62:4, 460–72.

Jacobs, J. (1993) *The Death and Life of Great American Cities*, New York: Modern Library.

Jones, L. (2000) *The Hermeneutics of Sacred Architecture: Experience, Interpretation, Comparison*, Cambridge, MA: Harvard University Press.

Kandinsky, W. (1994) *Complete Writings on Art*, ed. by K. C. Lindsay and P. Vergo, New York: Da Capo Press.

Kant, I. (1952) *The Critique of Judgement*, trans. by J. C. Meredith, Oxford: Oxford University Press.

Kidder, P. W. (1995) "Gadamer and the Platonic Eidos," *Philosophy Today* 39:1, 83–92.

Kierkegaard, S. (1992) *Concluding Unscientific Postscript to* Philosophical Fragments, Vol. I, trans. and ed. by H. V. Hong and E. H. Hong, Princeton, NJ: Princeton University Press.

Kolb, D. (1990) *Postmodern Sophistications: Philosophy, Architecture, and Tradition*, Chicago/London: The University of Chicago Press.

Lammi, W. (2008) *Gadamer and the Question of the Divine*, London/New York: Continuum.

Lawn, C. (2006) *Gadamer: A Guide for the Perplexed*, London/New York: Continuum.

Lawrence, F. G. (1984) "Language as Horizon?" in *The Beginning and the Beyond: Papers from the Gadamer and Voegelin Conferences*," ed. by F. G. Lawrence, Chico, CA: Scholars Press, 13–33.

—— (1990) "Baur's 'Conversation with Hans-Georg Gadamer' and 'Contribution to the Gadamer-Lonergan Discussion': A Reaction," *Method: Journal of Lonergan Studies* 8:2, 135–51.

—— (2000) "Ontology *of* and *as* Horizon: Gadamer's Rehabilitation of the Metaphysics of Light," *Revista Portuguesa de Filosofia*, 56:3/4, 389–420.

—— (2002) "The Hermeneutic Revolution and the Future of Theology," in A. Wierciński, ed., *Between the Human and the Divine: Philosophical and Theological Hermeneutics*, Toronto: The Hermeneutic Press.

—— (2009) "Lonergan's Retrieval of Thomas Aquinas's Conception of *Imago Dei*: The Trinitarian Analogy of Intelligible Emanations in God," *American Catholic Philosophical Quarterly* 83:3, 363–88.

Le Corbusier (1967) *The Radiant City: Elements of a Doctrine of Urbanism to*

Be Used as the Basis of Our Machine-Age Civilization, New York: Orion Press.

—— (1986) *Towards a New Architecture*, trans. by F. Etchells, New York: Dover Publications.

Leatherbarrow, D. (1993) *The Roots of Architectural Invention: Site, Enclosure, Materials*, Cambridge: Cambridge University Press.

—— (2000) *Uncommon Ground: Architecture, Technology, and Topography*, Cambridge, MA: MIT Press.

Macquarrie, J. (1955) *An Existentialist Theology: A Comparison of Heidegger and Bultmann*, New York: Macmillan.

Madison, G. B. (1989) "Gadamer/Derrida: The Hermeneutics of Irony and Power," in *Dialogue and Deconstruction: The Gadamer–Derrida Encounter*, ed. by D. P. Michelfelder and R. Palmer, Albany, NY: SUNY Press, 192–8.

Malpas, J., Arnswald, U., Kertscher, J., eds., (2002) *Gadamer's Century: Essays in Honor of Hans-Georg Gadamer*, Cambridge, MA: MIT Press.

Margerum, R. D. (2002) "Collaborative Planning: Building Consensus and Building a Distinct Model for Practice," *Journal of Planning Education and Research* 21:3, 237–53.

McCarthy, T. (1978) *The Critical Theory of Jürgen Habermas*, Cambridge, MA: MIT Press.

Michelfelder, D. P. and Palmer, R. E., eds. (1989) *Dialogue and Deconstruction: The Gadamer–Derrida Encounter*, Albany, NY: SUNY Press.

Mueller-Vollmer, Kurt, ed. (1998) *The Hermeneutics Reader: Texts of the German Tradition from the Enlightenment to the Present*, London/New York: Continuum.

Mugerauer, R. (1994) *Interpretations on Behalf of Place: Environmental Displacements and Alternative Responses*, Albany, NY: SUNY Press.

—— (1995) *Interpreting Environments: Tradition, Deconstruction, Hermeneutics*, Austin: University of Texas Press.

—— (2008) *Heidegger and Homecoming: The Leitmotif in the Later Writings*, Toronto: University of Toronto Press.

Nietzsche, F. (2006) *The Nietzsche Reader*, ed. by K. A. Pearson and D. Large, Oxford/Malden, MA: Blackwell Publishing.

Norberg-Schulz, C. (1975) *Meaning in Western Architecture*, New York: Praeger.

—— (1979) *Genius Loci: Towards a Phenomenology of Architecture*, New York: Rizzoli.

—— (1985) *The Concept of Dwelling: On the Way to Figurative Architecture*, New York: Electra/Rizzoli.

—— (2000) *Principles of Modern Architecture*, London: Andreas Papadakis Publisher.

Ormiston, G. L. and Schrift, A. D. (1990) *The Hermeneutic Tradition: From Ast to Ricoeur*, Albany, NY: SUNY Press.

Pallasmaa, J. (2009) *The Thinking Hand: Existential and Embodied Wisdom in Architecture*, Chichester: Wiley.

Palmer, R. (1969) *Hermeneutics: Interpretation Theory in Schleiermacher, Dilthey, Heidegger, and Gadamer*, Evanston, IL: Northwestern University Press.

Pérez-Gómez, A. (1983) *Architecture and the Crisis of Modern Science*, Cambridge, MA: MIT Press.

—— (2008) *Built upon Love: Architectural Longing after Ethics and Aesthetics*, Cambridge, MA: MIT Press.

Pérez-Gómez, A. and Pelletier, L. (1997) *Architectural Representation and the Perspective Hinge*, Cambridge, MA: MIT Press.

Plato (1997) *Complete Works*, ed. by J. M. Cooper, Indianapolis, IN: Hackett Publishing Co.

Rambow, R. and Seifert, J. (2006) "Paint Damage and Whiskering: How Use is Entering into Architectural Discourse," *GAM: Graz Architektur Magazin* 3: 10–29.

Richardson, W. J. (1974) *Heidegger: Through Phenomenology to Thought*, 3rd ed., The Hague: Martinus Nijhoff.

Ricoeur, P. (1981) *Hermeneutics and the Human Sciences*, ed. and trans. by J. B. Thompson, Cambridge: Cambridge University Press.

Rilke, R. M. (1982) "Archaic Torso of Apollo," *The Selected Poetry of Rainer Maria Rilke*, trans. by S. Mitchell, New York: Random House.

Risser, J. (1997) *Hermeneutics and the Voice of the Other: Re-Reading Gadamer's Philosophical Hermeneutics*, Albany, NY: SUNY Press.

—— (2000) "From Concept to Word: The Radicality of Philosophical Hermeneutics," *Continental Philosophy Review* 33:3, 309–25.

—— (2002) "*Phronesis* as Kairological Event," *Epoché* 7:1, 107–19.

—— (2007) "Saying and Hearing the Word: Language and the Experience of Meaning in Gadamer's Hermeneutics," *Journal of Ultimate Reality and Meaning*, 30:2, 146–55.

Rykwert, J. (1996) *The Dancing Column: On Order in Architecture*, Cambridge, MA: MIT Press.

Schleiermacher, F. (1998) *Hermeneutics and Criticism, and Other Writings*, ed. by A. Bowie, Cambridge: Cambridge University Press.

Schmidt, D. J. (1994) "Introduction" to H.-G. Gadamer, *Heidegger's Ways*, Albany, NY: SUNY Press.

—— (2008) "Heidegger, Gadamer, Klee: On Word and Image," *Internationales Jahrbuch für Hermeneutik*, 7: 191–209.

Schmidt, L. K. (2007) *Understanding Hermeneutics*, Stocksfield: Acumen Publishing Ltd.

Schön, D. (1983) *The Reflective Practitioner*, New York: Basic Books.

Shapiro, G. and Sica, A., eds., (1984) *Hermeneutics: Questions and Prospects*, Amherst, MA: University of Massachusetts Press, 54–65.

Sharr, A. (2006) *Heidegger's Hut*, Cambridge, MA: MIT Press.

—— (2007) *Heidegger for Architects*, London/New York: Routledge.

Sheehan, T. (1987) *Karl Rahner: The Philosophical Foundations*, Athens, OH: Ohio University Press.

—— (2001) "Geschichtlichkeit/Ereignis/Kehre," *Existentia (Meletai Sophias)* 11:3–4, 241–51.

Silverman, H. J., ed. (1991) *Gadamer and Hermeneutics*, London/New York: Routledge.

Simon, J. (1989) "Good Will to Understand and Will to Power: Remarks on an 'Improbable Debate'," in *Dialogue and Deconstruction: The Gadamer-Derrida Encounter*, ed. by D. P. Michelfelder and R. Palmer, Albany, NY: SUNY Press, 162–75.

Sirowy, B. (2010) *Phenomenological Concepts in Architecture: Towards a User-Oriented Practice*, doctoral thesis, Oslo: Oslo School of Architecture and Design.

Snodgrass, A. and Coyne, R. (2006) *Interpretation in Architecture: Design as a Way of Thinking*, London/New York, Routledge.

Tate, D. (2001) "The Speechless Image: Gadamer and the Claim of Modern Painting," *Philosophy Today* 45:1, 56–68.

—— (2002) "The Remembrance of Art," in A. Wierciński, ed., *Between the Human and the Divine: Philosophical and Theological Hermeneutics*, Toronto: The Hermeneutic Press.

—— (2008) "Transforming *Mimesis*: Gadamer's Retrieval of Aristotle's *Poetics*," *Epoché* 13:1, 185–208.

Till, J. (2009) *Architecture Depends*, Cambridge, MA: MIT Press.

Vesely, D. (2004) *Architecture in the Age of Divided Representation: The Question of Creativity in the Shadow of Production*, Cambridge, MA: MIT Press.

Vilhauer, M. (2010) *Gadamer's Ethics of Play: Hermeneutics and the Other*, Plymouth: Lexington Books.

Wachterhauser, B. R., ed. (1986) *Hermeneutics and Modern Philosophy*, Albany, NY: SUNY Press.

Warnke, G. (1987) *Gadamer: Hermeneutics, Tradition, and Reason*, Stanford, CA: Stanford University Press.

—— (2011) "The Hermeneutic Circle versus Dialogue," *The Review of Metaphysics*, 65:2, 91–112.

Weinsheimer, J. C. (1985) *Gadamer's Hermeneutics: A Reading of* Truth and Method, New Haven/London: Yale University Press.

Wierciński, A., ed. (2002a) *Between the Human and the Divine: Philosophical and Theological Hermeneutics*, Toronto: The Hermeneutic Press.

—— (2002b) "The Hermeneutic Retrieval of a Theological Insight: *Verbum Interius*," in A. Wierciński, ed., *Between the Human and the Divine: Philosophical and Theological Hermeneutics*, Toronto: The Hermeneutic Press.

Wigley, M. (1995) *The Architecture of Deconstruction: Derrida's Haunt*, Cambridge, MA: MIT Press.

Index

Aalto, Alvar 71
abstraction 25, 65, 82–3, 109
   in art 6, 9, 27–8, 61
   in urban planning 77–9
Anaxagoras 64
Aquinas, Thomas 107
architecture,
   aesthetic approach to 6–7, 18–20,
     32–3, 70, 103, 115–16
   decorative character of 20, 28–31,
     34
   design in *see* design
   ethical function of 7, 51, 69–72
   hermeneutic approach to 1, 5–10,
     28–30, 34, 50, 53–6, 88–9,
     91–2, 104, 120–3
   meaning in 2–3, 5–9, 19–20, 22–32,
     51–2, 71–2, 103–4, 119–21
   modern 7, 9, 19–20, 27–9, 50–6,
     58–63, 66, 69–72, 76–9, 103,
     115
   program in 29–31, 33, 59, 89
   sacred 54–6, 58–63, 119
   symbolism in 6–9, 19–20, 26–8,
     52, 54–6, 59, 63, 66, 69–71,
     103–4
   tradition and 6, 8–10, 34, 52–8,
     60–3, 69–71, 92–4, 105, 115
Aristotle 2, 11, 13–14, 66, 82–3, 112

Augustine 3, 107

Bacon, Francis 41, 67, 70
Baroque art 6, 19–20
Baudrillard, Jean 110
Barrie, Thomas 55–6, 119
Baumgarten, Alexander 20
beauty 6–7, 17–18, 30–1
Benveniste, Émile 110
Bernstein, Richard 115
Bultmann, Rudolf 119

Cobb, Henry N. 56
Coyne, Richard 54, 91, 115–16, 125
Christianity 3–4, 83, 100, 107,
   117–19

De Meuron, Pierre 26
De Saussure, Ferdinand 110
decoration *see* ornament
deconstruction 112–16, 125
Deleuze, Gilles 110
deliberative process 84–6,
Derrida, Jacques 15, 110–16
Descartes, René 41, 67, 70
design 32–3, 54, 66, 87, 89–92,
   115–16
dialectic 4, 21, 75
différance 113

Dilthey, Wilhelm 14, 49
dwelling 51–2, 68, 100, 103–5

Eames, Charles and Ray 90
effective history (*Wirkungsgeschichte*) 47
   consciousness of 48
Einstein, Albert 49
Enlightenment, The 4–5, 18, 41, 67, 118

Forester, John 86
Foucault, Michel 110
Freud, Sigmund 74–6, 110–11
Friedländer, Paul 11, 13
functionality 6, 9–10, 20, 30, 66, 71, 121
fusion of horizons (*Horizontverschmelzung*) 43–4, 86, 91–2, 94

Gadamer, Hans-Georg,
   influences on 12–15, 68
   life and education 11–12
Gallilei, Galileo 67
Gehry, Frank 33–6
Giedion, Sigfried 50
Graves, Michael 58
Gropius, Walter 7

Habermas, Jürgen 15, 72–5, 83, 102
Harries, Karsten 7, 13, 20, 50–1, 70–1, 103–4
Hegel, G. W. F. 10, 14, 68, 99
Heidegger, Martin 7, 10–13, 15–18, 29, 48, 51–3, 68–9, 97–108, 110–12, 117, 119
hermeneutic circle 3, 12, 91
hermeneutics,
   history of 13–14
   of suspicion 109–16
   of texts 2–3, 43–6, 83, 110–3
   philosophical 1–3, 5, 13–15, 18, 46–50, 69, 73–5, 81, 83–6, 117–19, 122
   universality of 2
   Romantic 3, 55, 88–9
Herzog, Jacques 24–6
historicism 48–9
history 9–10, 14, 34–7, 39, 45–9, 51–4, 60–1, 73, 85–6, 102, 108, 115, 120–1
Hobbes, Thomas 67
Holl, Steven 25–6, 32, 58–63, 87, 89–90, 109
Homer 118
horizon,
   idea of 38–50, 73, 84, 86, 88, 106, 117, 121
   and architecture 54–6, 89–92, 94
Hubbard, Bill Jr. 88–91, 109
Huizinga, Johan 17–18
Husserl, Edmund 10, 13, 68, 99

inner word 107, 119
International Style 56, 115

Jacobs, Jane 77–80
Johnson, Philip 58
Jones, Lindsay 55–6, 119

justice 14, 21, 65, 82–3, 94

Kahn, Louis 7, 71
Kandinsky, Wassily 28
Kant, Immanuel 10, 17–20, 22, 31, 68, 110, 117
Kierkegaard, Søren 14–15, 99
Kinnell, Galway 45

Lacan, Jacques 111
language 38–40, 43, 47, 65, 73–4, 100–2, 104–7, 111–15
Lawrence, Frederick 117
Le Corbusier (Charles-Édouard Jeanneret) 7, 29, 70–2, 76–7
Leatherbarrow, David 52–4, 104
Lévi-Strauss, Claude 110
Loos, Adolf 7, 29, 52–3, 70, 103
Luther, Martin 100, 117
Lyotard, Jean-François 110

method 4–5, 41–2, 66–9, 73–5, 91
Mies van der Rohe, Ludwig 7
Mockbee, Samuel 92–5
Moore, Charles 58
Moses, Robert 77–9
Mugerauer, Robert 13, 104–5

National Socialism 11, 51
natural science 4–5, 64–9, 73, 98, 110
naturalism 13, 64–9
New Urbanism 58, 80
Newton, Issac 4, 68
Nietzsche, Friedrich 15, 99, 109–12

Norberg-Schulz, Christian 7, 13, 51–2, 71, 104

ontology 97–109, 119, 121–2
ornament 6–7, 27–9, 32–3, 52–3, 70

Pei, I. M. 56
Pérez-Gómez, Alberto 7, 15, 69–72, 104
phenomenology 13, 54–5, 68, 99, 103–4
painting 23, 27–30, 115
Pallasmaa, Juhani 26, 32
Pantheon 25–6
Picasso, Pablo 9
Plato 2, 4, 13–14, 21–2, 64–6, 81–2, 112, 118
play (*Spiel*) 10, 17–19, 21–2, 28–9, 32–4, 44, 58, 101, 113, 120
practical skill (*techne*) 81, 83–4, 89, 122–3
practical wisdom (*phronesis*) 81–4, 95–6, 123
prejudice 40–3, 74–5, 111
postmodernism 7, 15, 32, 56–8, 110–12, 115–16
presentation (*Darstellung*) 23–4

Radiant City 76–8
Rahner, Karl 119
recognition (*Wiedererkennung*) 23–4, 28, 43, 48, 53–4, 56, 60–1, 72, 114–15
relativism 42, 48–50
Richardson, Henry Hobson 56–7

Ricoeur, Paul 109
Rilke, Rainer Maria 22, 24–5, 27
Risser, James 114–15, 125
Rural Studio 92–5

Safdie, Moshe 116
Schelling, F. W. J. 14
Schiller, Friedrich 19
Schön, Donald 91–2
Sirowy, Beata 92–5
Snodgrass, Adrian 54, 91, 115, 125
social science 72–6, 83
Socrates 13, 21–2, 49–50, 64–6, 81–3, 106, 118, 123
speechless image 22–8, 106–7
Stirling, James 32
Structuralism 110–12
subjectivity 45–6, 73, 99 100
  and art 17–22

Tate Modern Museum 26
time 32, 48, 69, 98, 107–9
toleration 42
Troeltsch, Ernst 49

truth 4–5, 10, 12, 14, 17–18, 21–3, 48–50, 101, 108, 114–15, 119–20

understanding 2–4, 18, 40–4, 55–6, 64–5, 75, 77, 84–9, 101, 106, 111, 113–14, 118–19, 121–2
universality 14, 49, 94–5, 109, 115
  in art 17–19, 24
  in history 14, 48–9
  of hermeneutics 2
University of Heidelberg 12, 31
urban planning 76–80, 86–7
Urban Renewal 77–80

Venturi, Robert 58
Vesely, Dalibor 2, 5, 7, 15, 19–20, 69, 125

Vitruvius (Marcus Vitruvius Pollio) 29, 69–70

Zumthor, Peter 26